AF581777

Processing and Technology of Dairy Products

Processing and Technology of Dairy Products

Special Issue Editors

Hilton Deeth
Phil Kelly

MDPI • Basel • Beijing • Wuhan • Barcelona • Belgrade • Manchester • Tokyo • Cluj • Tianjin

Special Issue Editors
Hilton Deeth
The University of Queensland
Australia

Phil Kelly
Teagasc Food Research Centre Moorepark
Ireland

Editorial Office
MDPI
St. Alban-Anlage 66
4052 Basel, Switzerland

This is a reprint of articles from the Special Issue published online in the open access journal *Foods* (ISSN 2304-8158) (available at: https://www.mdpi.com/journal/foods/special_issues/Processing_Technology_Dairy).

For citation purposes, cite each article independently as indicated on the article page online and as indicated below:

LastName, A.A.; LastName, B.B.; LastName, C.C. Article Title. *Journal Name* **Year**, *Article Number*, Page Range.

ISBN 978-3-03928-688-1 (Hbk)
ISBN 978-3-03928-689-8 (PDF)

Contents

About the Special Issue Editors

Hilton Deeth has long been involved in dairy research involving milk and a range of dairy products. His particular interests include milk lipids and lipolysis, milk proteins and proteolysis, and thermal and non-thermal technologies. He joined the University of Queensland, Australia, in 1995, where he taught food science and technology and supervised several postgraduate students and research projects. From 1996 to 2008, he was Director of the Centre for UHT Processing at the University of Queensland. He has published over 160 research papers and 35 book chapters. He co-authored the 2017 book entitled "High Temperature Processing of Milk and Milk Products" and co-edited the 2018 book on "Whey Proteins. From Milk to Medicine". He is now Emeritus Professor of Food Science at the University of Queensland and consults to the dairy industry.

Phil Kelly retired as a Senior Principal Research Officer following a long research career at Teagasc Food Research Centre at Moorepark in Ireland. A graduate originally in Dairy and Food Science from University College Cork, he later achieved Ph.D. and MBA postgraduate degrees from the same institution. During his career at Moorepark, he served for over 20 years in research management as Departmental Head in Dairy Technology, Food Ingredients and Food Processing & Functionality. His wide-ranging research interests included development and exploitation of novel process technologies especially developments in membrane separation for targeted enrichment and enhanced functionality of milk components for food formulation, infant nutrition and cheesemaking applications. He has researched and published on the interrelationships between spray drying parameters and resulting milk powder properties. He has supervised Ph.D. students; he is author/co-author of over 150 peer-reviewed research publications as well as book chapters, Guest Editor of Special Issues of the *International Dairy Journal*, and is the holder of intellectual property including a number of process-based patents.

He has had a long association with the International Dairy Federation: as elected member of IDF's Science & Programme Coordination Committee (SPCC) during 2015–2018; former National Secretary, Irish National Committee of IDF (1988–2016); Member of IDF's Commission B—Science, Engineering & Technology (1998–2000); and Chair of IDF's Standing Committee for Dairy Science and Technology (2007–2011). In 2016, he was the recipient of the IDF Award for outstanding contribution to progress in dairying worldwide.

 *foods*

Editorial

Processing and Technology of Dairy Products: A Special Issue

Hilton Deeth [1,*] and Phil Kelly [2]

1 School of Agriculture and Food Sciences, The University of Queensland, Brisbane 4072, Australia
2 Teagasc Food Research Centre Moorepark, Fermoy, P61 C996 Co. Cork, Ireland; philk51@hotmail.com
* Correspondence: h.deeth@uq.edu.au

Received: 27 February 2020; Accepted: 2 March 2020; Published: 3 March 2020

When this Special Issue was launched, we cast the net widely in terms of the subject matter we considered suitable for the papers. We stated that papers on "well-established unit operations such as heat treatments and membrane separation in addition to emerging technologies" would be welcomed. The seven papers accepted do, indeed, cover a range of topics including UHT milk, proteolytic digestion, membrane technologies, cheese and yogurt. Three papers [1–3] involve aspects of the beneficial uses of proteolytic enzymes, two [4,5] involve the use of membrane technology in cheese making, while two deal with the role of ingredients—raw milk in the UHT paper [6] and apricot fiber in the yogurt paper [7]—in product quality. All in all, the papers demonstrate the breadth of ongoing research for an industry based on just one raw material, milk.

Each submission explores innovative approaches by the respective authors in their quest to push the boundaries of scientific and technological understanding. Some examples are illustrated below: Chamberland et al. [4] address the question of whether one should chose a 0.1 µm pore size MF or 10 kDa molecular weight cut-off ultrafiltration (UF) membranes for cheese milk standardization. The authors found that the UF, rather than the MF membrane, scored better in terms of lower running (energy and membrane) costs. In a related study, the sensory quality of hard, high-cooked cheese processed from milk, preconcentrated 1.9 fold by reverse osmosis, is shown by Taivosalo et al. [5] to be largely unaffected. Heat stability represents an important field of study in dairy science, and in this issue, readers have the opportunity to consider the approach of Karlsson et al. [6], who undertook a full factorial designed study on the role of key milk components on the stability of UHT milk. Crude preparations of apricot fiber (Karaca et al. [7]) were demonstrated as a novel ingredient with the capacity to confer functional benefits during yogurt processing.

As editors of this special issue, we find it appropriate to reflect on how well the scientific originality of the reviewed manuscripts scored against sustainability criteria. The two membrane-based papers concerned with either protein standardization [4] or milk pre-concentration [5] impact cheesemaking efficiency directly through yield improvements, shorter manufacturing processes and increased manufacturing capacity (without the need for extra cheese vat capacity). Chamberland et al. [4] go one step further by differentiating between closely matched permeating UF and MF polymeric membranes in favor of UF, because of its lower energy usage and membrane replacement costs. Protein hydrolysis is not typically associated with an opportunity to fractionate whey protein, except Sáez et al. [3] identified an opportunity during a particular set of incubation conditions in which the breakdown of α-lactalbumin (α-la) could be delayed. Before putting in place a strategy to recover undigested α-la, Sáez et al. [3] identified a number of shortcomings during application of a range of non-thermal and thermal methods of inactivating the enzyme-containing hydrolysate. Chief among these was the unexpected amount of heat-induced aggregation taking place among peptides and undigested protein in the whey protein hydrolysates which ruled out subsequent fractionation efforts. Suddenly, what was perceived to be a very elegant, sustainable, dual enzymatic hydrolysis/fractionation process was ground to a halt and is pending the next stage of development. The positive interaction when a plant-based ingredient is shown to be functional in yogurt [7} is evidence of how synergies may be harnessed when a holistic approach is adopted via food ingredient combinations, i.e., plant and dairy can co-exist in formulated foods where they can be enjoyed for both the pleasure of eating as well as the health benefits that they bring to the consumer.

It is important to recognize the valuable dissemination contribution that ***foods*** is making through this Special Issue: Processing and Technology of Dairy Products, given its achievement of a high impact factor, its commitment to a rapid turnaround of peer-reviewed manuscripts before publication, and its accessibility to a wide audience.

Conflicts of Interest: The authors declare no conflict of interest.

References

1. Estrada, O.; Ariño, A.; Juan, T. Salt Distribution in Raw Sheep Milk Cheese during Ripening and the Effect on Proteolysis and Lipolysis. *Foods* **2019**, *8*, 100. [CrossRef] [PubMed]
2. Gutiérrez-Méndez, N.; Balderrama-Carmona, A.; García-Sandoval, S.E.; Ramírez-Vigil, P.; Leal-Ramos, M.Y.; García-Triana, A. Proteolysis and Rheological Properties of Cream Cheese Made with a Plant-Derived Coagulant from Solanum elaeagnifolium. *Foods* **2019**, *8*, 44. [CrossRef] [PubMed]
3. Sáez, L.; Murphy, E.; FitzGerald, R.J.; Kelly, P. Exploring the Use of a Modified High-Temperature, Short-Time Continuous Heat Exchanger with Extended Holding Time (HTST-EHT) for Thermal Inactivation of Trypsin Following Selective Enzymatic Hydrolysis of the β-Lactoglobulin Fraction in Whey Protein Isolate. *Foods* **2019**, *8*, 367.
4. Chamberland, J.; Mercier-Bouchard, D.; Dussault-Chouinard, I.; Benoit, S.; Doyen, A.; Britten, M.; Pouliot, Y. On the Use of Ultrafiltration or Microfiltration Polymeric Spiral-Wound Membranes for Cheesemilk Standardization: Impact on Process Efficiency. *Foods* **2019**, *8*, 198. [CrossRef] [PubMed]
5. Taivosalo, A.; Kriščiunaite, T.; Stulova, I.; Part, N.; Rosend, J.; Sõrmus, A.; Vilu, R. Ripening of Hard Cheese Produced from Milk Concentrated by Reverse Osmosis. *Foods* **2019**, *8*, 165. [CrossRef] [PubMed]
6. Karlsson, M.A.; Lundh, Å.; Innings, F.; Höjer, A.; Wikström, M.; Langton, M. The Effect of Calcium, Citrate, and Urea on the Stability of Ultra-High Temperature Treated Milk: A Full Factorial Designed Study. *Foods* **2019**, *8*, 418. [CrossRef] [PubMed]
7. Karaca, O.B.; Güzeler, N.; Tangüler, H.; Yaşar, K.; Akın, M.B. Effects of Apricot Fibre on the Physicochemical Characteristics, the Sensory Properties and Bacterial Viability of Nonfat Probiotic Yoghurts. *Foods* **2019**, *8*, 33. [CrossRef] [PubMed]

Article

Effects of Apricot Fibre on the Physicochemical Characteristics, the Sensory Properties and Bacterial Viability of Nonfat Probiotic Yoghurts

Oya Berkay Karaca [1,*], Nuray Güzeler [2], Hasan Tangüler [3], Kurban Yaşar [4] and Mutlu Buket Akın [5]

1 Karatas School of Tourism and Hotel Management, Cukurova University, 01903 Adana, Turkey
2 Agricultural Faculty, Department of Food Engineering, Cukurova University, 01330 Adana, Turkey; nguzeler10@gmail.com
3 Faculty of Engineering, Department of Food Engineering, Nigde University, 51245 Nigde, Turkey; htanguler@nigde.edu.tr
4 Department of Food Engineering, Osmaniye Korkut Ata University, 80000 Osmaniye, Turkey; kurbanyasar@osmaniye.edu.tr
5 Faculty of Engineering, Department of Food Engineering, Harran University, 63100 Şanlıurfa, Turkey; mutluakin@harran.edu.tr
* Correspondence: obkaraca@cu.edu.tr; Tel.: +90-322-696-8401 (ext. 130)

Received: 25 December 2018; Accepted: 15 January 2019; Published: 18 January 2019

Abstract: In this study, the physical, chemical, rheological, and microbiological characteristics and the sensory properties of nonfat probiotic yoghurt produced at two different concentrations of apricot fibre (1% and 2%, *w*/*v*) and three different types of probiotic culture (*Lactobacillus* (*L.*) *acidophilus* LA-5, *Bifidobacterium animalis* subsp. *lactis* BB-12 (*Bifidobacterium* BB-12), and their mixtures) were investigated. As the fibre content increased, the rheological, structural, and sensory properties of probiotic yoghurt were negatively affected, while counts of *L. delbrueckii* subsp. *bulgaricus*, *L. acidophilus* LA-5, and *Bifidobacterium* BB-12 increased. When all the results were evaluated, the best results were obtained by using *L. acidophilus* LA-5 as probiotic culture and adding 1% (*w*/*v*) apricot fibre.

Keywords: apricot fibre; *Bifidobacterium* BB-12; *L. acidophilus* LA-5; lactic and acetic acids; probiotic yoghurt

1. Introduction

Consumers across the world are becoming more interested in foods with health-promoting features as they gain more awareness of the links between food and health. Among functional foods, products containing probiotics are showing promising trends worldwide [1]. Probiotics such as *Lactobacillus* and *Bifidobacterium* spp. are bacterial members of the human gut microbiota that exert several beneficial effects on human health and well-being through the production of short-chain fatty acids, which improves the intestinal microbial balance, resulting in the inhibiting bacterial pathogens, reducing colon cancer risk, stimulating the immune system, and lowering serum cholesterol levels [2]. In order to produce therapeutic benefits, a suggested minimum level for probiotic bacteria in fermented milk is above 10^6 cfu mL^{-1} [3]. Several factors are responsible for the viability of these organisms, e.g., the strains used, growth conditions, antagonism among cultures present, storage time and temperature, initial counts, hydrogen peroxide and oxygen contents in the medium, and the amount of organic acids in the product [4]. Considerable studies have been conducted to stimulate the growth of probiotic bacteria during yoghurt fermentation and to improve their survival until the use-by date, by supplementing yoghurt milk with growth factors such as vitamin-enriched protein hydrolysate, amino nitrogen whey protein concentrate and cysteine [5]. A recent approach is to incorporate prebiotic substrates to support the growth and activity of probiotics [6,7].

Apricot is a rich source of sugars, fibres, minerals, bioactive phytochemicals, and vitamins like A, C, thiamine, riboflavin, niacin, and pantothenic acid. Among the phytochemicals, phenolics, carotenoids, and antioxidants are important for their biological value [8]. The aim of this study was to investigate the effects of apricot fibre (AF) on the overall quality and viable bacteria counts in nonfat probiotic yoghurt in order to set up the best formulation in the supplementation of food-grade fibre. For this purpose, nonfat probiotic yoghurts were manufactured by adding different rates of AF and individual and mixture cultures of *Lactobacillus acidophilus* LA-5 and *Bifidobacterium animalis* subsp. *lactis* BB-12.

2. Materials and Methods

2.1. Materials

Raw cow's milk used in the experiments was obtained from the Animal Husbandry section of the Agricultural Faculty. Lyophilised starter cultures (coded FYS11, Marshall, France) containing *Streptococcus* (*Str.*) *thermophilus* and *Lactobacillus* (*L.*) *delbrueckii* subsp. *bulgaricus* were used as starter culture. In addition, lyophilised *L. acidophilus* LA-5 and *Bifidobacterium animalis* subsp. *lactis* BB-12 cultures were obtained from CHR-Hansen Company (Hørsholm, Denmark). All bacteria were maintained on de Man, Rogosa and Sharpe (MRS) agar slants. Apricots were used as dietary fibre source, and they were obtained from Apricot Research Institute in the first week of July, when they had attained enough maturity.

2.2. Methods

2.2.1. Apricot Fibre Production

Fresh apricots were washed, cut lengthwise, and their kernel was removed. The apricot pieces were added to water containing citric acid (1%, w/v) to avoid any browning. They were taken from the water containing citric acid, and their water was removed. They were placed in freezer bags in the refrigerator. Then, they were dried in a freeze dryer (Ilshin, FD-8512, ilShin Biobase Europe B.V., Kryptonstraat 33, Netherlands) at −70 °C (condenser temperature) with 5 mTorr of pressure and a total time cycle of 48 h (freeze-drying time). Dried apricots were broken into powder in a blender (Heidolph Diax 900, Merck KGaA, Darmstadt, Germany) and sieved to remove large pieces. The obtained apricot fibre (moisture 5%, fat 0.40%, protein 4.00%, ash 3.80%, sugar 90.40%, cellulose 4.00%) was stored in closed plastic containers in a freezer at −20 °C.

2.2.2. The Production of Yoghurt Containing Apricot Fibre Using Probiotic Culture

The fat content in the cow's milk that was brought to the dairy technology laboratory was adjusted to 0.1% (v/v) using a cream separator (Elecrem, Vanves, France). Then, the milk was divided into seven parts, and milk powder and apricot fibre were added at the different levels given in Table 1. After blending the milk with milk powder and apricot fibre, the mixtures were separately homogenised using an Ultra Turrax blender (IKA, Merck, Germany) at 14,000 rpm until all ingredients were dissolved. Then, the homogenates were pasteurised at 85 °C for 5 min and cooled to 45 ± 1 °C. Starter culture (3%, v/v) and probiotic culture were added at a rate of 3% (about 10^6 cfu mL^{-1}; 1:1) to the cooled milks and filled into plastic yoghurt cups (200 mL). The samples were incubated in a Medcenter incubator (Friocell, Planegg/München, Germany) at 43 ± 1 °C until pH 4.7. At the end of incubation, all yoghurt samples were stored at refrigerator temperature (4 ± 1 °C) for 20 days. Yoghurt production was performed in triplicate. They were analysed after 1, 10, and 20 days of storage.

2.2.3. Chemical Analyses

Analysis of Apricot Fibre

Total solids, protein, ash, and total amount of dietary fibre, fat contents, and sugar by soluble solids were determined according to Association of Official Analytical Chemists (AOAC) [9].

Chemical Analysis of Yoghurts

Total solids, fat, titratable acidity, protein, and ash [9] were measured. The pH values were measured using a digital pH meter (WTW, Wielheim, Germany). Lactic acid and acetic acid were analysed by an HPLC (Shimadzu LC-20AD, Shimadzu Corporation, Tokyo, Japan) using an Aminex HPX-87H column (Bio-Rad, Hercules, California, USA) at 50 °C. The eluent was 5 mmol L^{-1} H_2SO_4 in high-purity water at a flow rate of 0.6 mL min^{-1}. Lactic acid and acetic acid amounts were calculated from a UV detector [10]. Standards (Merck, Darmstadt, Germany) were used to determine the concentration of organic acids. Replicates for all analytical determinations were carried out in duplicate.

Table 1. Microorganisms and additives used in the production of yoghurt of different types.

Yog	AF (%)	MP (%)	Bacteria
A	0	6	yoghurt bacteria
B	1	5	yoghurt bacteria + *Lactobacillus acidophilus* LA-5
C	1	5	yoghurt bacteria + *Bifidobacterium* BB-12
D	1	5	yoghurt bacteria + *Lactobacillus acidophilus* LA-5 + *Bifidobacterium* BB-12
E	2	4	yoghurt bacteria + *Lactobacillus acidophilus* LA-5
F	2	4	yoghurt bacteria + *Bifidobacterium* BB-12
G	2	4	yoghurt bacteria + *Lactobacillus acidophilus* LA-5+*Bifidobacterium* BB-12

Yog: Yoghurt; AF: Apricot fibre; MP: Milk powder; Yoghurt bacteria: *Streptococcus thermophilus* and *Lactobacillus delbrueckii* subsp. *bulgaricus*.

Physical Measurements

Gel firmness was measured in the experimental yoghurts using a penetrometer model SUR BERLIN PNR 6 (Berlin, Germany) with a 15 g conical (45°) probe. Results were expressed as 1/10 millimetres of the penetration within 5 s. The viscosity values of the samples were determined using a Brookfield viscometer (model DV-II + Pro, Brookfield Engineering Laboratories, Middleboro, MA, USA) at 4 °C with a spindle (S64) rotation of 100 rpm and by applying a single constant shear rate (0.05 s^{-1}). The readings were recorded at the 15th second of the measurement. The measurements were taken three times for each yoghurt sample, and the readings were recorded as centipoises. Water holding capacity (WHC) was determined using the centrifuge method with a modified procedure [11]. For this purpose, 5 g native yoghurt (NY) was centrifuged at 483× *g* for 30 min at 10 °C. After centrifugation, the supernatant was removed, and whey expelled (WE) was weighed and expressed as a percentage of yoghurt weight. Whey separation was considered as the amount of drained liquid (g) per twenty-five grams of sample. Each sample was weighed on a filter paper no. 589/2 (12.5 cm, 0.00009 g) placed on top of a funnel. The drainage time and temperature were 120 min and +4 °C, respectively [12]. The colour of the yoghurt samples was measured with a Minolta Chroma Meter CR-100 (Minolta, Osaka, Japan). L* (brightness, 100 = white, 0 = black), a* (+, red; −, green) and b* (+, yellow; −, blue) values were measured.

Microbiological Analysis of Yoghurts

For the counts of yoghurt and probiotic bacteria, 10 g of yoghurt samples were homogenised in 90 ml of peptone water (0.1% peptone) in a Stomacher 400 (Type BA7021, Seward Medical, Londan, UK) for at least 2 min. Samples were serially diluted in peptone water and spread inoculated (0.1 mL) onto plates. Plate counts of *Str. thermophilus* and *L. delbrueckii* subsp. *bulgaricus* were performed in

M17 agar and MRS agar (Merck, Istanbul, Turkey), respectively. Incubations were conducted at 37 °C for 2 days (aerobically) and at 43 °C for 3 days (anaerobically), respectively. *L. acidophilus* LA-5 and *Bifidobacterium* BB-12 were enumerated on MRS-Sorbitol Agar (1% sorbitol) and MRS-NNLP Agar (100 mg neomycine sulphate, 15 mg nalidixic acid and 3 g LiCl), respectively [13,14]. The plates were incubated at 37 °C for 3 days (anaerobically). Anaerobic conditions were created using Anaerocult A sachets (Merck). Plates containing 20–200 colonies were counted, and the results were expressed as colony-forming units per gram (cfu g^{-1}) of sample.

Sensory Characteristics

Sensory characteristics of the yoghurt samples were evaluated by a panel of ten expert members from the Laboratory of Milk and Dairy Products at Cukurova University according to a 0–5-point scale [15].

Statistical Analysis

Data were calculated for statistical significance by one-way analysis of variance (ANOVA). Means compared by Duncan's test for statistical analysis were carried out. All analyses were made in triplicate and performed on the 1st, 10th, and 20th days of storage, except composition of milk and yoghurts.

3. Results and Discussion

3.1. Composition of Milk and Yoghurt

Titratable acidity (0.16 ± 0.01% LA), pH value (6.70 ± 0.00), and dry matter (8.81 ± 0.11%), fat (0.10 ± 0.00%), ash (0.74 ± 0.05%), and protein (3.33 ± 0.18%) contents of the nonfat milk used in the production of yoghurt were determined. The pH values, titratable acidity, and composition values of yoghurts were determined on the first day of storage and are given in Table 2. In the performed statistical evaluation, differences between titratable acidity values of yoghurts were found to be significant ($p < 0.05$). On the first day of storage, pH values, dry matter, protein, fat, and ash rates of yoghurts proved to be similar, whereas significant differences between titratable acidity values were found ($p < 0.05$). When the addition of apricot fibre rate increased so that dry matter did not change, milk powder and the total addition of ingredient rate was adjusted 6% (w/v). Garcia-Perez et al. [16], specified to no effect from the addition of orange fibre of yoghurt composition.

Table 2. Physicochemical composition of AF added nonfat yoghurts ($n = 3$).

	pH	Titratable Acidity (LA %)	Dry Matter (%)	Protein (%)	Fat (%)	Ash (%)
Milk	6.70 ± 0.00	0.160 ± 0.01	8.81 ± 0.11	3.33 ± 0.18	0.10 ± 0.00	0.74 ± 0.05
A	4.70 ± 0.07 ^{a}	1.138 ± 0.14 ^{a}	13.83 ± 0.40 ^{a}	5.64 ± 0.71 ^{a}	0.10 ± 0.00 ^{a}	1.28 ± 0.09 ^{a}
B	4.75 ± 0.06 ^{a}	1.056 ± 0.09 ^{a,b,c}	13.79 ± 0.18 ^{a}	5.16 ± 0.92 ^{a}	0.10 ± 0.00 ^{a}	1.21 ± 0.04 ^{a}
C	4.81 ± 0.05 ^{a}	1.002 ± 0.05 ^{b,c}	13.71 ± 0.22 ^{a}	5.04 ± 0.27 ^{a}	0.10 ± 0.00 ^{a}	1.22 ± 0.07 ^{a}
D	4.81 ± 0.08 ^{a}	1.020 ± 0.05 ^{b,c}	13.76 ± 0.12 ^{a}	4.18 ± 0.54 ^{a}	0.10 ± 0.00 ^{a}	1.25 ± 0.08 ^{a}
E	4.83 ± 0.13 ^{a}	1.023 ± 0.07 ^{b,c}	13.75 ± 0.13 ^{a}	4.59 ± 0.34 ^{a}	0.10 ± 0.00 ^{a}	1.33 ± 0.25 ^{a}
F	4.87 ± 0.17 ^{a}	0.971 ± 0.07 ^{c}	13.69 ± 0.18 ^{a}	4.87 ± 0.23 ^{a}	0.10 ± 0.00 ^{a}	1.15 ± 0.03 ^{a}
G	4.80 ±0.15 ^{a}	1.090 ±0.08 ^{a,b}	13.60 ± 0.23 ^{a}	4.37 ± 0.59 ^{a}	0.10 ± 0.00 ^{a}	1.14 ± 0.02 ^{a}

^{a,b,c} Means in the same column followed by different letters were significantly different ($p < 0.05$).

3.2. Changes in Chemical Properties of Probiotic Set Yoghurts during Storage

The pH, titratable acidity, and lactic and acetic acid changes of yoghurts during the storage period are given in Table 3. While pH values decreased considerably, titratable acidity values increased ($p > 0.05$). Additionally, yoghurts E, F, G, which had the highest added AF rate, had the highest values of titratable acidity after yoghurt A in general ($p < 0.01$). Lario et al. [17] specified that the addition of 1% (w/v) orange fibre to yoghurts caused a decrease in pH values of yoghurts and improved structural properties.

Table 3. Physicochemical properties of nonfat probiotic set yoghurts during storage ($n = 3$).

Day	Yoghurts A	Yoghurts B	Yoghurts C	Yoghurts D	Yoghurts E	Yoghurts F	Yoghurts G
				pH			
1	$4.70 \pm 0.08^{A,a}$	$4.75 \pm .06^{A,a}$	$4.81 \pm 0.05^{A,a}$	$4.81 \pm 0.09^{A,a}$	$4.83 \pm 0.15^{A,a}$	$4.87 \pm 0.20^{A,a}$	$4.80 \pm 0.17^{A,a}$
10	$4.54 \pm 0.12^{AB,a}$	$4.51 \pm 0.21^{A,a}$	$4.63 \pm 0.17^{AB,a}$	$4.60 \pm 0.15^{AB,a}$	$4.58 \pm 0.09^{B,a}$	$4.61 \pm 0.04^{AB,a}$	$4.52 \pm 0.10^{B,a}$
20	$4.37 \pm 0.11^{B,a}$	$4.41 \pm 0.15^{B,a}$	$4.45 \pm 0.14^{B,a}$	$4.47 \pm 0.15^{B,a}$	$4.42 \pm 0.07^{B,a}$	$4.41 \pm 0.13^{B,a}$	$4.37 \pm 0.06^{B,a}$
				Titratable acidity (LA%)			
1	$1.138 \pm 0.16^{A,a}$	$1.056 \pm 0.11^{A,a}$	$1.002 \pm 0.06^{B,a}$	$1.020 \pm 0.06^{B,a}$	$1.023 \pm 0.08^{B,a}$	$0.971 \pm 0.08^{C,a}$	$1.090 \pm 0.09^{A,a}$
10	$1.192 \pm 0.02^{A,a}$	$1.182 \pm 0.04^{A,a}$	$1.150 \pm 0.06^{A,a}$	$1.165 \pm 0.07^{A,a}$	$1.197 \pm 0.09^{A,a}$	$1.102 \pm 0.05^{B,a}$	$1.125 \pm 0.04^{A,a}$
20	$1.272 \pm 0.07^{A,a}$	$1.153 \pm 0.03^{A,a}$	$1.243 \pm 0.06^{A,a}$	$1.233 \pm 0.07^{A,a}$	$1.265 \pm 0.010^{A,a}$	$1.243 \pm 0.04^{A,a}$	$1.250 \pm 0.10^{A,a}$
				Lactic acid (g/L)			
1	$16.59 \pm 0.04^{C,a}$	$12.27 \pm 0.08^{C,b}$	$11.48 \pm 0.23^{C,d}$	$12.66 \pm 0.21^{C,b}$	$11.70 \pm 0.15^{B,cd}$	$12.94 \pm 0.17^{C,b}$	$12.03 \pm 0.35^{C,c}$
10	$18.70 \pm 0.11^{B,a}$	$14.38 \pm 0.10^{B,c}$	$14.42 \pm 0.21^{B,c}$	$14.41 \pm 0.18^{B,c}$	$12.31 \pm 0.47^{B,d}$	$14.42 \pm 0.20^{B,c}$	$14.51 \pm 0.27^{B,c}$
20	$19.48 \pm 0.06^{A,a}$	$15.15 \pm 1.40^{A,bc}$	$16.42 \pm 0.27^{A,b}$	$15.53 \pm 0.51^{A,bc}$	$14.32 \pm 1.03^{A,c}$	$16.43 \pm 0.36^{A,b}$	$16.45 \pm 0.08^{A,b}$
				Acetic acid (g/L)			
1	$0.792 \pm 0.00^{B,a}$	$0.595 \pm 0.0 8^{A,c}$	$0.518 \pm 0.01^{C,d}$	$0.697 \pm 0.01^{C,b}$	$0.439 \pm 0.02^{C,e}$	$0.708 \pm 0.00^{C,b}$	$0.574 \pm 0.00^{C,cd}$
10	$0.851 \pm 0.02^{A,a}$	$0.671 \pm 0.05^{A,b}$	$0.813 \pm 0.03^{B,a}$	$0.776 \pm 0.03^{B,a}$	$0.550 \pm 0.03^{B,c}$	$0.840 \pm 0.02^{B,a}$	$0.821 \pm 0.07^{B,a}$
20	$0.887 \pm 0.04^{A,b}$	$0.755 \pm 0.05^{A,c}$	$1.282 \pm 0.11^{A,a}$	$1.218 \pm 0.03^{A,a}$	$0.615 \pm 0.00^{A,d}$	$0.957 \pm 0.04^{A,b}$	$0.954 \pm 0.00^{A,b}$

[a,b,c,d,e] Means in the same row followed by different letters were significantly different ($p < 0.05$). [A,B,C] Means in the same column followed by different letters were significantly different ($p < 0.05$).

In dairy products, lactic acid is one of major compounds of lactose degradation due to the lactic acid bacterial fermentation. During the fermentation of milk, depending on the microorganisms involved in the medium, lactic acid is produced via the glycolysis pathway while lactic and acetic acids are formed via the pentose phosphate pathway [18]. Lactic acid, which acts on milk protein, gives to yoghurt its texture and its characteristic sensory properties [19]. The highest amount of lactic acid was found in yoghurt A (without AF) on the first day of storage. The yoghurts produced with apricot fibre had lower levels of lactic and acetic acid than the control sample did on the first day of storage. In addition, lactic and acetic acid amounts decreased with increasing apricot fibre addition, and the effects of the addition of apricot fibre on the amount of lactic acid and acetic acid were significant ($p < 0.01$).

The amounts of lactic acid and acetic acid of yoghurts significantly increased while pH decreased significantly in yoghurts during the storage period ($p < 0.05$). Similar results were obtained by Ong et al. [20], who have stated that acetic acid amounts in cheddar produced by probiotic cultures increased during the storage period. *Str. thermophilus* and *L. delbrueckii* subsp. *bulgaricus* use lactose homofermentatively to produce lactic acid, whilst *Bifidobacterium* spp. produces lactic acid and acetic acid due to their heterofermentative nature by fermenting the same sugar [21]. As expected in the present study, lactic and acetic acid amounts in yoghurt samples produced with yoghurt starter cultures and *Bifidobacterium* BB-12 were higher than those in samples produced with yoghurt starter cultures and *L. acidophilus* LA-5 after 20 days of storage.

3.3. Changes in Rheological and Structural Properties of Probiotic Set Yoghurts

The changes in gel firmness, whey separation, water holding capacity, viscosity values during the storage period are given in Table 4. The effect of using apricot fibre on gel firmness values of yoghurts was found to be significant ($p < 0.05$). As the added AF rate increased, the titratable acidity values decreased, consequently decreasing gel firmness values, so yoghurts had a softer body. It was determined that gel firmness values of yoghurts with 2% (w/v) AF added were considerably different from those of yoghurts with 1% (w/v) AF on the 1st and 10th days of storage ($p < 0.05$). It is thought that the weakening of the gel structure is due to the addition of fibre instead of milk powder. Lario et al. [17] expressed that the rheological properties of yoghurt change related to added fibre ratios. The gel firmness values of yoghurts decreased during the storage period, and then it was observed that this decrease was significant in yoghurts A, E, and F ($p < 0.05$). The decrease of gel firmness values during the storage period arose from hydration of casein micelles of clot [22].

The least whey separation rate was in yoghurt C and the highest was in yoghurt G on the first day of storage. This alignment did not change on the 10th and 20th days of storage, that is, the highest whey separation rate was 2% in yoghurt G. In the case of the increasing rate of added AF and decreasing quantity of milk powder, it was determined that whey separation rates of yoghurts were increased ($p < 0.01$). A decrease in the amount of protein from the milk powder may lead to an increase in serum separation. Lario et al. [17] expressed that the changes of rheological properties of yoghurt are related to added fibre ratios. They reported that on the condition of 1% (w/v) orange fibre added to yoghurt, the quantity of whey separation was reduced and there were improved structural properties. In addition, Ferrandez Garcia and Gregor [23] informed that viscosity increased by rice and corn fibre addition to yoghurt, but it did not increase by sugar beet and soybean addition to yoghurt. The effect of different culture types on this feature was not statistically significant ($p > 0.05$). It was found that the whey separation values of all yoghurts decreased during the storage period, and this decrease was significant for yoghurts B, D, F, and G ($p < 0.05$).

Table 4. Rheological and structural properties of nonfat probiotic set yoghurts (n = 3).

Day	Yoghurts A	Yoghurts B	Yoghurts C	Yoghurts D	Yoghurts E	Yoghurts F	Yoghurts G
Gel firmness (mm/5 sn)							
1	196 ± 3.79 ^{A,e}	196 ± 1.73 ^{A,de}	203 ± 4.73 ^{A,cd}	200 ± 3.79 ^{A,cde}	206 ± 5.03 ^{A,bc}	214 ± 2.52 ^{A,a}	211 ± 2.89 ^{A,ab}
10	195 ± 5.00 ^{A,bcd}	194 ± 6.08 ^{A,cd}	192 ± 5.03 ^{A,d}	192 ± 3.79 ^{A,cd}	201 ± 2.52 ^{AB,abc}	203 ± 5.51 ^{B,ab}	206 ± 3.22 ^{A,a}
20	183 ± 4.51 ^{B,d}	188 ± 4.04 ^{A,cd}	191 ± 5.51 ^{A,bc}	192 ± 4.16 ^{A,bc}	196 ± 1.73 ^{B,b}	195 ± 3.79 ^{B,bc}	204 ± 4.16 ^{A,a}
Whey separation (%)							
1	18.03 ± 2.08 $^{A\ bc}$	17.51 ± 1.96 ^{A,bc}	15.43 ± 2.83 ^{A,c}	20.04 ± 1.82 ^{A,b}	24.11 ± 3.32 ^{A,a}	23.95 ± 0.24 ^{A,a}	24.64 ± 1.79 ^{A,a}
10	15.19 ± 0.88 ^{A,c}	16.89 ± 1.44 ^{A,c}	14.73 ± 2.76 ^{A,c}	17.28 ± 0.20 ^{B,bc}	20.88 ± 2.69 ^{A,a}	20.21 ± 1.42 ^{B,ab}	22.56 ± 1.49 ^{AB,a}
20	14.43 ± 1.85 ^{A,c}	13.41 ± 0.20 ^{B,c}	14.18 ± 0.30 ^{A,c}	16.33 ± 0.20 ^{B,b}	17.86 ± 1.26 ^{A,b}	19.79 ± 0.21 ^{B,a}	21.00 ± 0.50 ^{B,a}
Water holding capacity (%)							
1	69.97 ± 2.35 ^{A,a}	70.00 ± 2.21 ^{A,a}	70.70 ± 1.99 ^{A,a}	71.87 ± 1.63 ^{A,a}	70.83 ± 4.46 ^{A,a}	72.13 ± 3.51 ^{A,a}	72.83 ± 2.73 ^{A,a}
10	65.33 ± 3.41 ^{A,a}	67.50 ± 2.44 ^{A,a}	67.97 ± 2.57 ^{A,a}	70.07 ± 2.84 ^{A,a}	69.27 ± 1.72 ^{A,a}	70.33 ± 1.83 ^{A,a}	71.77 ± 2.87 ^{A,a}
20	63.20 ± 3.73 ^{A,b}	66.70 ± 1.21 ^{A,ab}	67.43 ± 2.54 ^{A,ab}	68.40 ± 2.71 ^{A,a}	69.30 ± 2.43 ^{A,a}	69.90 ± 1.74 ^{A,a}	69.90 ± 1.35 ^{A,a}
Viscosity (100 cp)							
1	2001 ± 77 ^{C,b}	2150 ± 39 ^{C,a}	1917 ± 37 ^{C,bc}	1948 ± 82 ^{C,bc}	2035 ± 21 ^{B,b}	2009 ± 71 ^{C,b}	1876 ± 84 ^{C,c}
10	2876 ± 40 ^{B,a}	2437 ± 48 ^{B,b}	2422 ± 10 ^{B,b}	2409 ± 71 ^{B,b}	2355 ± 33 ^{A,b}	2213 ± 58 ^{B,c}	2151 ± 58 ^{B,c}
20	3048 ± 11 ^{A,a}	2906 ± 35 ^{A,b}	2837 ± 21 ^{A,c}	2595 ± 12 ^{A,d}	2408 ± 66 ^{A,e}	2391 ± 38 ^{A,e}	2351 ± 50 ^{A,e}

^{a,b,c,d,e} Means in the same row followed by different letters were significantly different ($p < 0.05$); ^{A,B,C} Means in the same column followed by different letters were significantly different ($p < 0.05$).

It was found that the water holding capacity of yoghurts had values close to each other on the first day of storage, and these rates decreased, for all yoghurts during storage ($p > 0.05$). It was determined that the water holding capacity of yoghurts was not affected by the addition of AF and by different probiotic cultures on the 1st and 10th days of storage ($p > 0.05$), and the water holding capacity of 2% (*w*/*v*) AF added yoghurts was affected considerably on the 20th day of storage ($p < 0.05$). Yoghurts F and G had the highest water holding capacity values. It can be stated that the result may be due to the high water-binding capacity of the fibres. Güler-Akın et al. [7] reported that the addition of apple fibre caused an increase in the water holding capacity of yoghurts.

Viscosity values were significantly affected by the added AF rate ($p < 0.01$). The viscosity values of yoghurts D and G produced with *L. acidophilus* LA-5 + *Bifidobacterium* BB-12 mixture culture were lower than those of other yoghurts produced with single culture. It was determined that the viscosity values of yoghurts increased considerably during the storage period ($p < 0.01$), and viscosity values of 1% (*w*/*v*) AF added yoghurts were higher than those of 2% (*w*/*v*) AF added yoghurts at this increase. The protein content which had an influence on the viscosity decreased with decreasing milk powder ratio in yoghurts; as a result, the viscosities of the yoghurts decreased. The highest value of viscosity was determined in the control yoghurt on the last day of storage ($p < 0.01$). Çayır [24] determined that the viscosity values of apricot puree added yoghurts increased during the storage period.

3.4. Changes in Colour Characteristics of Probiotic Set Yoghurts during Storage

L*, a*, b* values of yoghurts and their changes in time during the storage period are given in Table 5. The values of L* and a* were not influenced by culture types ($p > 0.05$). The L* values of yoghurts ranged between 87.09–90.32 on the first day of storage. The differences between L* values of yoghurts were determined to be very significant at all terms of storage time ($p < 0.01$). It was determined that the whiteness of yoghurt A (which has no AF added) was highest and L* values of yoghurts B, C, and D (which have 1% (*w*/*v*) AF added) were higher than yoghurts E, F, G (which have 2% (*w*/*v*) AF added), that is, when the AF increased, the values of L* decreased. The L* values of yoghurts at the end of storage decreased according to values on the first day, and this change was very significant in yoghurts B, E, and F ($p < 0.05$). Sanz et al. [25] specified that the brightness of yoghurts decreased by the addition of asparagus fibre, and yellow-green values increased. Garcia-Perez et al. [26] determined that orange fibre addition affected the colour of yoghurt: The L* value decreased, and the a* and b* values increased.

On the first day of storage, the a* values of yoghurts decreased (from −3.98 and −4.84) as the added AF rate increased. The difference between a* values of yoghurt A and the others is significant at all periods of storage ($p < 0.01$). It was found that the a* values of yoghurts F and G, which have the highest added AF rate, were lower than those of the control yoghurt on the 20th day of storage ($p < 0.05$). The a* values of yoghurts increased on the 20th day of storage compared to values on the first day of storage, so yoghurts had a greener colour by the end of storage. As the AF increased, the red colour increased in yoghurts as well. The storage period effect was not found to be significant on the a* values of yoghurt ($p > 0.05$).

Table 5. Colour properties in nonfat probiotic set yoghurts during storage ($n = 3$).

Day	Yoghurts A	Yoghurts B	Yoghurts C	Yoghurts D	Yoghurts E	Yoghurts F	Yoghurts G
				L^*			
1	$90.32 \pm 0.72^{A,a}$	$89.22 \pm 0.55^{A,b}$	$89.46 \pm 0.40^{A,ab}$	$89.46 \pm 0.56^{A,ab}$	$87.74 \pm 0.14^{A,c}$	$87.76 \pm 0.47^{A,c}$	$87.09 \pm 0.68^{A,c}$
10	$90.12 \pm 0.74^{B,a}$	$87.37 \pm 0.48^{B,cd}$	$88.80 \pm 0.61^{A,b}$	$88.01 \pm 0.83^{A,bc}$	$86.20 \pm 0.95^{B,de}$	$85.34 \pm 0.47^{B,e}$	$85.44 \pm 0.10^{A,de}$
20	$90.29 \pm 0.47^{A,a}$	$88.33 \pm 0.44^{AB,bc}$	$87.76 \pm 0.87^{A,bcd}$	$88.62 \pm 0.90^{A,b}$	$86.83 \pm 0.14^{AB,d}$	$87.26 \pm 0.90^{A,cd}$	$86.52 \pm 0.85A^{d}$
				a^*			
1	$-4.84 \pm 0.34^{A,a}$	$-4.26 \pm 0.31^{A,b}$	$-4.34 \pm 0.23^{A,b}$	$-4.30 \pm 0.15^{A,b}$	$-4.01 \pm 0.20^{A,b}$	$-3.98 \pm 0.25^{A,b}$	$-4.00 \pm 0.23^{A,b}$
10	$-4.83 \pm 0.27^{A,a}$	$-4.07 \pm 0.01^{A,b}$	$-4.01 \pm 0.02^{A,b}$	$-4.06 \pm 0.24^{A,b}$	$-3.77 \pm 0.14^{A,bc}$	$-3.70 \pm 0.22^{A,c}$	$-3.80 \pm 0.22^{A,bc}$
20	$-4.70 \pm 0.12^{A,a}$	$-4.27 \pm 0.14^{A,b}$	$-4.14 \pm 0.11^{A,bc}$	$-3.95 \pm 0.16^{A,cd}$	$-3.90 \pm 0.14^{A,cd}$	$-3.87 \pm 0.17^{A,d}$	$-3.88 \pm 0.10^{A,d}$
				b^*			
1	$8.29 \pm 0.89^{A,d}$	$11.25 \pm 0.83^{A,bc}$	$11.06 \pm 0.41^{A,c}$	$11.08 \pm 0.08^{A,c}$	$12.34 \pm 0.86^{B,ab}$	$12.52 \pm 0.25^{B,a}$	$12.63 \pm 0.73^{B,a}$
10	$8.37 \pm 0.65^{A,c}$	$11.44 \pm 0.15^{A,b}$	$11.39 \pm 0.34^{A,b}$	$11.11 \pm 0.79^{A,b}$	$12.91 \pm 0.47^{B,a}$	$12.80 \pm 0.81^{B,a}$	$12.54 \pm 0.65^{B,a}$
20	$8.60 \pm 1.06^{A,d}$	$12.83 \pm 0.78^{A,b}$	$12.06 \pm 0.94^{A,bc}$	$11.48 \pm 0.12^{A,c}$	$14.18 \pm 0.04^{A,a}$	$14.18 \pm 0.02^{A,a}$	$14.74 \pm 0.84^{A,a}$

[a,b,c,d,e] Means in the same row followed by different letters were significantly different ($p < 0.05$); [A,B,C] Means in the same column followed by different letters were significantly different ($p < 0.05$).

As the added AF rate increased, b* values increased significantly. Two percent (*w*/*v*) AF added yoghurts had higher b* values on the first day of storage and during the storage period. The b* values of yoghurts increased during the storage period, and the added AF rate had a significant effect on the b* values of yoghurts during the storage period ($p < 0.01$). It was determined that changes in b* values were significant in the 2% (*w*/*v*) AF added yoghurts E, F, and G during the storage period ($p < 0.01$).

3.5. Bacterial Counts of Probiotic Set Yoghurts during Storage

Viable bacterial counts of yoghurt samples during the storage period are shown in Table 6. In spite of the sample F having the highest *Str. thermophilus* counts, added AF rate and using a different probiotic culture did not affect the number of *Str. thermophilus* ($p > 0.05$). The counts of *Str. thermophilus* decreased slowly during the storage period, and storage period had significant effect on the *Str. thermophilus* counts.

Usage of a different probiotic culture had no effect on the *L. delbrueckii* subsp. *bulgaricus* counts of yoghurt ($p > 0.05$). On the other hand, the addition of different rates of AF improved the viability of *L. delbrueckii* subsp. *bulgaricus* ($p < 0.01$). *L. delbrueckii* subsp. *bulgaricus* counts increased with increasing added AF rate. There was an approximately 1-log cycle decrease in the counts of *L. delbrueckii* subsp. *bulgaricus* at the end of storage ($p < 0.01$).

The culture type had significantly affected *L. acidophilus* LA-5 counts of samples ($p < 0.01$). The *L. acidophilus* LA-5 counts of yoghurt which was inoculated with a mixture of *L. acidophilus* LA-5 and *Bifidobacterium* BB-12 were higher than those of the yoghurt inoculated with individual *L. acidophilus* LA-5. Studies showed *Bifidobacterium* spp. and *L. acidophilus* LA-5 strains live in excellent symbiosis, and their counts were increased when they were inoculated together [27,28]. The addition of different rates of AF also improved the survival of *L. acidophilus* LA-5 ($p < 0.01$). The counts of *L. acidophilus* LA-5 increased as added AF rate increased due to the possible prebiotic effects of AF. Similar results were reported for many fibres by the other authors [29,30]. The counts of *L. acidophilus* LA-5 decreased during the storage period. The most important factors affecting the viability of *L. acidophilus* LA-5 are acidity and hydrogen peroxide [13]. The acidity of the samples increased during the storage period. During the storage period, although the viable counts of *L. acidophilus* LA-5 dropped in all samples, the counts in all probiotic yoghurts were found to be above the threshold for therapeutic minimum (10^6–10^7 cfu g^{-1}).

Table 6. The changes of viable bacteria counts of nonfat probiotic yoghurts during storage (log cfu g^{-1}) ($n = 3$).

Yoghurts	Storage Period (Day)	*S. thermophilus*	*L. delbrueckii* subsp. *bulgaricus*	*L. acidophilus* LA-5	*Bifidobacterium* BB-12
A	1	8.34 ± 0.08 [a]	8.07 ± 0.01 [abcde]	-	-
	10	7.87 ± 0.04 [bc]	7.91 ± 0.06 [de]	-	-
	20	7.73 ± 0.01 [c]	7.28 ± 0.13 [fg]	-	-
B	1	8.36 ± 0.11 [a]	8.12 ± 0.04 [abcd]	7.91 ± 0.05 [a]	-
	10	7.76 ± 0.06 [bc]	7.83 ± 0.03 [e]	7.35 ± 0.03 [d]	-
	20	7.72 ± 0.01 [c]	7.26 ± 0.19 [fg]	6.34 ± 0.03 [g]	-
C	1	8.33 ± 0.07 [a]	8.15 ± 0.10 [abcd]	-	7.40 ± 0.07 [a]
	10	7.85 ± 0.05 [bc]	7.91 ± 0.05 [de]	-	6.31 ± 0.11 [b]
	20	7.77 ± 0.02 [c]	7.33 ± 0.16 [fg]	-	5.26 ± 0.02 [d]
D	1	8.33 ± 0.06 [a]	8.19 ± 0.04 [ab]	7.96 ± 0.03 [a]	7.47 ± 0.07 [a]
	10	7.86 ± 0.06 [bc]	7.43 ± 0.05 [f]	7.50 ± 0.06 [bc]	6.21 ± 0.10 [b]
	20	7.72 ± 0.04 [c]	7.13 ± 0.12 [g]	6.56 ± 0.04 [f]	5.43 ± 0.09 [cd]
E	1	8.35 ± 0.14 [a]	8.22 ± 0.08 [a]	7.96 ± 0.01 [a]	-
	10	7.92 ± 0.01 [b]	7.91 ± 0.04 [de]	7.41 ± 0.05 [cd]	-
	20	7.73 ± 0.07 [c]	7.42 ± 0.14 [f]	6.65 ± 0.04 [ef]	-
F	1	8.40 ± 0.10 [a]	8.09 ± 0.04 [abcd]	-	7.49 ± 0.06 [a]
	10	7.90 ± 0.01 [bc]	7.93 ± 0.03 [cde]	-	6.40 ± 0.25 [b]
	20	7.81 ± 0.05 [c]	7.32 ± 0.17 [fg]	-	5.64 ± 0.05 [c]
G	1	8.33 ± 0.12 [a]	8.17 ± 0.10 [abc]	7.98 ± 0.02 [a]	7.49 ± 0.05 [a]
	10	7.90 ± 0.03 [bc]	7.94 ± 0.01 [bcde]	7.60 ± 0.04 [b]	6.45 ± 0.09 [b]
	20	7.72 ± 0.06 [c]	7.28 ± 0.10 [fg]	6.75 ± 0.03 [c]	5.68 ± 0.06 [c]

[a,b,c,d,e,f,g] Different letters indicate significant differences among the samples ($p < 0.01$).

The use of probiotic culture as individual or mixture had no effect statistically on the *Bifidobacterium* BB-12 ($p > 0.05$). However, Gomes and Malcata [31] reported that growth of *L. acidophilus* LA-5 caused to diminish the redox potential in the samples and so improved the counts of *Bifidobacterium* BB-12. On the other hand, the effect of different rates of AF was significant on the viability of *Bifidobacterium* BB-12 ($p < 0.01$). The counts of *Bifidobacterium* BB-12 increased as added AF rate increased ($p < 0.01$), but this increase was not enough to reach above 10^6–10^7 cfu g^{-1}. *Bifidobacterium* BB-12 counts decreased during the storage period ($p < 0.01$), which could be attributed to antagonistic relationships between yoghurt bacteria and probiotic strains. At the end of storage, the counts of *Bifidobacterium* BB-12 decreased to approximately 1-log cycle in all probiotic yoghurts, and the counts were found to be below the threshold for therapeutic minimum (10^6–10^7 cfu g^{-1}). Authors reported that some fibres, such as xylo-oligosaccharides, lactulose, rafinose, inulin, and citrus fibre had a prebiotic effect on the *Bifidobacterium* spp. [29].

3.6. Changes in Sensory Properties of Probiotic Set Yoghurt during Storage

The scores recorded for appearance, consistency (by spoon and mouth), taste, and total sensory properties of yoghurts are given in Figure 1. The highest appearance score on the first day of storage was determined in the control yoghurt, and the lowest appearance scores were received by yoghurts F and G on the last day of storage ($p > 0.05$). Different fibre ratios did not affect appearance scores at all periods of storage ($p > 0.05$). While the consistency (with spoon) scores of yoghurts A and C were found to be the highest, yoghurts E and G had the lowest scores. It was determined that as the addition of AF increased and milk powder decreased, the consistency of scores of yoghurts with spoon decreased at all periods of storage ($p < 0.01$). This result can be explained by the protein content and the effect of the protein on the gel structure and is parallel to the results of gel firmness. On the first day of storage, while there was no difference between yoghurts, on the 10th and 20th days in the control yoghurt and the 1% (w/v) AF added yoghurt, consistency (with spoon) values were similar ($p > 0.05$) except for 2% (w/v) AF added yoghurts ($p < 0.01$).

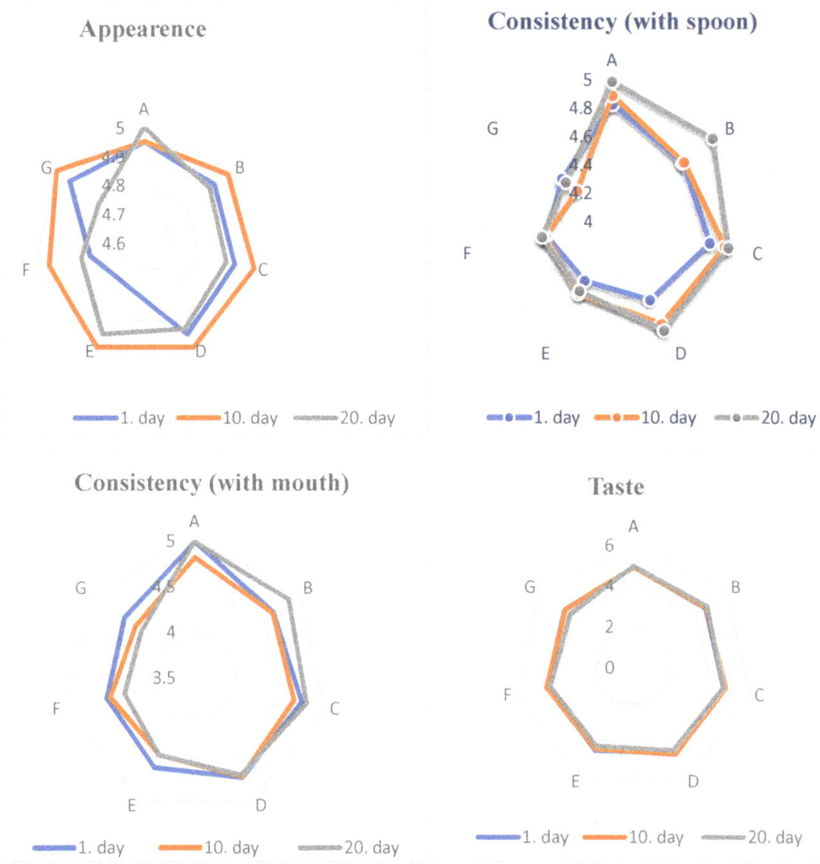

Figure 1. Sensory properties of probiotic set yoghurt during storage.

As the added AF rate increased, the consistency score in mouth decreased. It was established that consistency scores with the mouth decreased according to first day values in yoghurts during the storage period. The effect of added AF rate was found to be significant on the difference between consistency scores of yoghurts ($p < 0.05$). The effect of storage time on the consistency scores with mouth was not significant except for yoghurt B ($p > 0.05$). The effect of added AF rate was not significant on the odour properties of yoghurts ($p > 0.05$), and as added AF rate increased, odour scores decreased.

The highest score of taste was yoghurt A without AF, and the lowest score was yoghurt G. While the difference between the flavour properties of yoghurt on the first day of storage was not found to be important, the control yoghurt and the 2% (w/v) AF added yoghurts E, F, and G were found to be significant and considerably different on the 10th and 20th days ($p < 0.01$). Generally, flavour scores of 1% (w/v) AF added yoghurts were the highest scores after those of yoghurt A. Additionally, 2% (w/v) AF added decreased the scores. During all periods of storage, yoghurt A had the highest total sensory score, and following this, the 1% (w/v) AF added yoghurts on the last day of storage did not change this order. It was determined that the effect of storage time was not significant on the odour, flavour, and total sensory scores of yoghurts ($p > 0.05$). The difference between the total sensory scores of yoghurts were very significant on the 10th and 20th days of storage ($p < 0.05$). Yoghurts containing individual cultures were found to have higher scores than those of their mixtures. Fernandez-Garcia and McGregor [23] specified that bamboo, wheat, and inulin fibre added yoghurts had sensory properties close to the control yoghurt. They informed that the sensory and rheological properties of apple fibre added yoghurt were different from those of control yoghurt. Additionally, Staffolo et al. [32] specified that bamboo, wheat, and inulin fibre added yoghurts had sensory properties close to control yoghurt despite apple fibre addition being different significant.

4. Conclusions

The results of the analysis indicate that as the added AF rate increases, there is an increase in whey separation, gel firmness, water holding capacity, a* and b* values, *L. delbrueckii* subsp. *bulgaricus*, *L. acidophilus* LA-5, *Bifidobacterium* BB-12 counts of yoghurts, and a decrease in lactic and acetic acid amounts, the titratable acidity, viscosity, L* values, consistency, odour and flavour scores. The viscosity values of yoghurts D and G produced with *L. acidophilus* LA-5 + *Bifidobacterium* BB-12 mixture culture were lower than those of other yoghurts produced with single culture. The quantities of lactic and acetic acid using *Bifidobacterium* BB-12 probiotic culture were found to be higher on the first day and during the storage period. In the sensory evaluation, yoghurt A had the highest total sensory score, and following this, during all periods of storage, the 1% (w/v) AF added yoghurts. The usage of *L. acidophilus* LA-5 culture in yoghurts acquired more appreciation according to *Bifidobacterium* BB-12 and mixed cultures addition of yoghurt. According to the results, yoghurts containing individual cultures had higher scores than those of their mixtures and using *L. acidophilus* LA-5 as the probiotic culture and 5% milk powder +1% (w/v) AF addition for nonfat probiotic yoghurt production can be suggested.

Author Contributions: Examined the literature, O.B.K.; Methodology, O.B.K., K.Y., H.T., and M.B.A.; Performed the Analysis, O.B.K., M.B.A., H.T., and K.Y.; Project Administration, N.G. and O.B.K.; Writing—Original Draft Preparation, O.B.K., M.B.A., and H.T.; Review and Editing, Proof-Reading-Service.com; O.B.K. is the supervisor of article submission.

Funding: This research was funded by TÜBİTAK (Scientific & Technical Research Council of Turkey) grant number TOGTAG/TARP-107O421.

Acknowledgments: This research was financed by Project TOGTAG/TARP-107O421 from TÜBİTAK (Scientific & Technical Research Council of Turkey).

Conflicts of Interest: The authors declare no conflict of interest. The founding sponsors had no role in the design of the study; in the collection, analyses, or interpretation of data; in the writing of the manuscript, and in the decision to publish the results.

References

1. Paseephol, T.; Sherkat, F. Probiotic stability of yoghurts containing jerusalem artichoke inulins during refrigerated storage. *J. Funct. Foods* **2009**, *1*, 311–318. [CrossRef]
2. Saarela, M.; Lahteenmaki, L.; Crittenden, R.; Salminen, S.; Mattila-Sandholm, T. Gut bacteria and health foods: The European perspective. *Int. J. Food Microbiol.* **2002**, *78*, 99–117. [CrossRef]
3. Rodrigues, D.; Sousa, S.; Rocha-Santos, T.; Silva, J.P.; Sousa Lobo, J.M.; Costa, P.; Amaral, M.H.; Pintado, M.M.; Gomes, A.M.; Malcata, F.X.; et al. Influence of L-cysteine, oxygen and relative humidity upon survival throughout storage of probiotic bacteria in whey protein-based microcapsules. *Int. Dairy J.* **2011**, *21*, 869–876. [CrossRef]
4. Shah, N.P. Probiotic bacteria: Selective enumeration and survival in dairy products. *J. Dairy Sci.* **2000**, *83*, 894–907. [CrossRef]
5. Amatayakul, T.; Halmos, A.L.; Sherkat, F.; Shah, N.P. Physical characteristics of yoghurts made using exopolysaccharide-producing starter cultures and varying casein to whey protein ratios. *Int. Dairy J.* **2006**, *16*, 40–51. [CrossRef]

6. Srisuvor, N.; Chinprahast, N.; Prakitchaiwattana, C.; Subhimaros, S. Effects of inulin and polydextrose on physicochemical and sensory properties of low-fat set yoghurt with probiotic-cultured banana puree. *LWT-Food Sci. Technol.* **2013**, *51*, 30–36. [CrossRef]
7. Güler-Akın, M.B.; Ferliarslan, I.; Akın, M.S. Apricot probiotic drinking yoghurt supplied with inulin and oat fiber. *Adv. Microbiol.* **2016**, *6*, 999–1009. [CrossRef]
8. Ali, S.; Masud, T.; Abbasi, K.S. Physico-chemical characteristics of apricot (*Prunus armeniaca* L.) grown in northern areas of Pakistan. *Sci. Hortic.* **2011**, *30*, 386–392. [CrossRef]
9. AOAC. *Official Methods of Analysis of AOAC International*, 18th ed.; Association of Official Analytical Chemists: Gaithersburgs, MD, USA, 2006.
10. Tanguler, H.; Erten, H. Occurrence, and growth of lactic acid bacteria species during the fermentation of shalgam (salgam), a traditional Turkish fermented beverage. *LWT-Food Sci. Technol.* **2012**, *46*, 36–41. [CrossRef]
11. Remeuf, F.; Mohammed, S.; Sodini, I.; Tissier, J.P. Preliminary observations on the effects of milk fortification and heating on microstructure and physical properties of stirred yoghurt. *Int. Dairy J.* **2003**, *13*, 773–782. [CrossRef]
12. Tamime, A.Y.; Barrantes, E.; Sword, A.M. The effects of starch based fat substitutes on the microstructure of set-style yogurt made from reconstituted skimmed milk powder. *J. Soc. Dairy Technol.* **1996**, *49*, 1–10. [CrossRef]
13. Dave, R.I.; Shah, N.P. Viability of yogurt and probiotic bacteria in yogurts made from commercial starter cultures. *Int. Dairy J.* **1997**, *7*, 31–41. [CrossRef]
14. Gustaw, W.; Kordowska-Wiater, M.; Koziol, J. The influence of selected prebiotics on the growth of lactic acid bacteria for bio-yoghurt production. *Acta Sci. Pol. Technol. Aliment.* **2011**, *10*, 455–466. [PubMed]
15. Anonymous. *TS 1330 Yoğurt Standardı*; Türk Standartları Enstitüsü: Ankara, Turkey, 1989; 10p.
16. Garcia-Perez, F.J.; Sendra, E.; Lario, Y.; Fernandez-Lopez, J.; Sayas-Barbera, E.; Perez-Alvarez, J.A. Rheology of orange fiber enriched yogurt. *Milchwissenschaft* **2006**, *61*, 55–59.
17. Lario, Y.; Sendra, E.; Garcia-Perez, C.; Fuentes, E.; Sayas-Barbera, J.; Fernandez-Lopez, J.; Perez-Alvarez, J.A. Preparation of high dietary fiber powder from lemon juice by-products. *Innov. Food Sci. Emerg. Technol.* **2004**, *5*, 113–117. [CrossRef]
18. Shima, A.; Salina, H.; Masniza, M.; Atiqah, A. Viability of lactic acid bacteria in homemade yogurt containing sago starch oligosaccharides. *Int. J. Basic Appl. Sci.* **2012**, *12*, 58–62.
19. Serafeimidou, A.; Zlatanos, S.; Laskaridis, K.; Sagredos, A. Chemical characteristics, fatty acid composition and conjugated linoleic acid (CLA) content of traditional Greek yogurts. *Food Chem.* **2012**, *134*, 1839–1846. [CrossRef]
20. Ong, L.; Henriksson, A.; Shah, N.P. Proteolytic pattern and organic acid profiles of probiotic cheddar cheese as influenced by probiotic strains of *L. acidophilus*, *Lb. paracasei*, *Lb. casei* or *Bifidobacterium* sp. *Int. Dairy J.* **2007**, *17*, 67–78. [CrossRef]
21. Wu, S.C.; Wang, F.J.; Pan, C.L. Growth and survival of lactic acid bacteria during the fermentation and storage of seaweed oligosaccharides solution. *J. Mar. Sci. Technol.* **2007**, *15*, 104–114.
22. Tamime, A.Y.; Robinson, R.K. *Yoghurt Science and Technology*, 2nd ed.; Woodhead Publishing Limited: Cambridge, UK, 2000; 605p.
23. Fernandez-Garcia, E.; Mcgregor, J.U. Fortification of sweetened plain yogurt with ınsoluble dietary fiber. *Food Res. Int.* **1997**, *204*, 433–437. [CrossRef]
24. Çayır, M.S. *Probiyotik Kültür Kullanılarak Üretilen Kayısı Katkılı Yoğurtların Bazı Özellikleri*; Ç. Ü. Fen Bilimleri Enstitüsü: Adana, Turkey, 2007; 57p.
25. Sanz, T.; Salvador, A.; Jimenez, A.; Fiszman, S.M. Yogurt enrichment with functional asparagus fibre, effect of fibre extraction method on rheological properties, colour and sensory acceptance. *Eur. Food Res. Technol.* **2008**, *227*, 1515–1521. [CrossRef]
26. Garcia-Perez, F.J.; Lario, Y.; Fernandez-Lopez, J.; Sayas-Barbera, E.; Perez-Alvare, J.A.; Sendra, E. Effect of orange fiber addition on yogurt color during fermantation and cold storage. *J. Sci. Food Agric.* **2005**, *30*, 257–263. [CrossRef]
27. Helland, M.H.; Wicklund, T.; Narvhus, J.A. Growth and metabolism of selected strains of probiotic bacteria in milk and water-based cereal puddings. *Int. Dairy J.* **2004**, *14*, 957–965. [CrossRef]

28. Oliveira, R.P.S.; Florence, A.C.R.; Silva, R.C.; Perego, P.; Converti, A.; Gioielli, L.A.; Oliveira, M.N. Effect of different prebiotics on the fermentation kinetics, probiotic survival and fatty acids profiles in nonfat symbiotic fermented milk. *Int. J. Food Microbiol.* **2009**, *128*, 467–472. [CrossRef] [PubMed]
29. Sendra, E.; Fayos, P.; Lario, Y.; Fernandez-Lopez, J.; Sayas-Barbera, E.; Perze-Alvarez, J.A. Incorporation of citrus fibers in fermented milk containing probiotic bacteria. *Food Microbiol.* **2008**, *25*, 13–21. [CrossRef] [PubMed]
30. Oliveira, R.P.S.; Florence, R.C.; Perego, P.; Converti, A.; Oliveira, M.N. Effect of inulin on growth and acidification performance of different probiotic bacteria in co-cultures and mixed culture with *Streptococcus thermophilus*. *J. Food Eng.* **2009**, *91*, 133–139. [CrossRef]
31. Gomes, A.M.P.; Malcata, F.X. *Bifidobacterium* spp. and *Lactobacillus acidophilus*: Biological, biochemical, technological and therapeutical properties relevant for use as probiotics. *Trends Food Sci. Technol.* **1999**, *10*, 139–157. [CrossRef]
32. Staffolo, M.D.; Bertola, N.; Martino, M.; Bevilacqua, A. Influence of dietary fiber addition on sensory and rheological properties of yogurt. *Int. Dairy J.* **2004**, *14*, 263–268. [CrossRef]

Article

Proteolysis and Rheological Properties of Cream Cheese Made with a Plant-Derived Coagulant from *Solanum elaeagnifolium*

Néstor Gutiérrez-Méndez *, Alejandro Balderrama-Carmona, Socorro E. García-Sandoval, Pamela Ramírez-Vigil, Martha Y. Leal-Ramos and Antonio García-Triana

Facultad de Ciencias Químicas, Universidad Autónoma de Chihuahua, Chihuahua 31125, Mexico; alexbcarmona@gmail.com (A.B.-C.); eliizabeth.gs@gmail.com (S.E.G.-S.); pamelarvigil@hotmail.com (P.R.-V.); yarelyleal@hotmail.com (M.Y.L.-R.); triana.antonio@gmail.com (A.G.-T.)

* Correspondence: ngutierrez@uach.mx; Tel.: +52(614)-23-66-000 (ext. 4245)

Received: 20 December 2018; Accepted: 24 January 2019; Published: 30 January 2019

Abstract: Cream cheese is a fresh acid-curd cheese with pH values of 4.5–4.8. Some manufacturers add a small volume of rennet at the beginning of milk fermentation to improve the texture of the cream cheese. However, there is no information about the effect that proteases other than chymosin-like plant-derived proteases may have on cream cheese manufacture. This work aimed to describe some proteolytic features of the protease extracted from fruits of *Solanum elaeagnifolium* Cavanilles and to assess the impact that this plant coagulant has on the viscoelastic properties of cream cheeses. Results showed that caseins were not hydrolyzed extensively by this plant-derived coagulant. In consequence, the ratio of milk clotting units (*U*) to proteolytic activity (*U-Tyr*) was higher (1184.4 *U*/*U-Tyr*) than reported for other plant proteases. The plant coagulant modified neither yield nor composition of cream cheeses, but viscoelastic properties did. Cream cheeses made with chymosin had a loss tangent value ($tan\ \delta = 0.257$) higher than observed in cheeses made with 0.8 mL of plant-derived coagulant per liter ($tan\ \delta = 0.239$). It is likely that casein fragments released by the plant-derived coagulant improve the interaction of protein during the formation of acid curds, leading to an increase in the viscoelastic properties of cream cheese.

Keywords: plant-derived coagulant; *Solanum elaeagnifolium*; acid-rennet-curd cheese; cream cheese; proteolysis; texture

---

1. Introduction

Cream cheese is a fresh acid-curd cheese fermented with lactic acid bacteria at pH values of ~4.5–4.8 [1,2]. During acidification of milk, colloidal calcium phosphate (CCP) in the casein micelles are solubilized [3]. If the pH decreases below 5.0, most of CCP dissolves and casein micelles become dissociated [2]. When the pH in the milk reaches a value ~4.6, the negative surface charge on casein micelles will be reduced enough to induce the formation of a gel structure [4]. The firmness of acid-milk gels can be improved by either a preheat treatment of milk [2,5] or by a small addition of rennet [3]. Heat treatment of milk above 69 °C denatures the whey proteins (β-lactoglobulin (β-lg) and α-lactalbumin (α-la)) which associate with caseins [2]. Aggregation of β-lg with κ-casein reduces the net repulsive charge among caseins, enhancing the protein–protein interactions and thus the gel firmness [6]. On the other hand, addition of a small amount of rennet (chymosin) at the beginning of fermentation induces a coarser network of proteins that enhances the firmness of milk gels [3]. Chymosin hydrolyzes caseins (mostly κ-casein) generating *para*-κ-casein and a glycomacropeptide. However, it is not clear enough how the release of these casein fragments improves the interactions among proteins during formation of acid-milk gels [7]. Furthermore, there is no information about the impact of proteases other than chymosin in the formation of acid-enzymatic gels.

For centuries, the coagulant used for cheese making was obtained from calf stomach (calf rennet). Nowadays, calf rennet alone cannot cover the current demand for cheese-making coagulating enzymes. Recombinant *Bos taurus* chymosin and microbial-origin coagulants are alternative milk clotting enzymes widely used in the dairy industry nowadays [8]. Plant-derived coagulants have also been studied as possible substitutes for calf rennet, though in reality only a few are used for commercial cheese-making [9]. Most plant proteases are not suitable milk clotting agents because of their excessive proteolytic nature or their low ratio of milk clotting activity to proteolytic activity [8,9]. However, some plant-derived proteases are successfully used in the manufacture of commercial cheeses, for instance, plant-derived coagulants obtained from *Cynara cardunculus* L. and certain *Solanum* plants. The aspartic proteases from *Cynara* sp. are used to produce a large variety of cheeses in the Mediterranean, West Africa, and southern European countries [9]. Meanwhile, plant-derived coagulants from *Solanum dubium* Fresen and *Solanum elaeagnofolium* Cavanilles are used in the manufacture of commercial cheeses in Sudan [10] and Mexico [11].

In the northeast of Mexico, the fruits of *S. elaeagnifolium* (Figure 1) have been used for decades in the manufacture of asadero cheese, a pasta-filata-type cheese [12]. The use of this plant for asadero cheese manufacture was first described between 1916 [13] and 1924 [14]. Unfortunately, the protease from *S. elaeagnifolium* is lesser-known and not as well studied as proteases from *Cynara* sp. or *S. dubium*. Some of the most interesting features of protease from *S. elaeagnifolium* is its high ratio of milk clotting activity to proteolytic activity [11]. Additionally, this plant-derived coagulant can induce the formation of stable milk-gels and curds with a soft texture, particularly at mild-acidic conditions [11,12]. In consequence, the protease from fruits of *S. elaeagnifolium* could be used in the manufacture of acid-enzymatic curd cheeses with a soft texture. This work aimed to describe some proteolytic features of the protease extracted from ripening fruits of *S. elaeagnifolium*, as well as to assess the usage of this plant-derived coagulant in the manufacture of cream cheese.

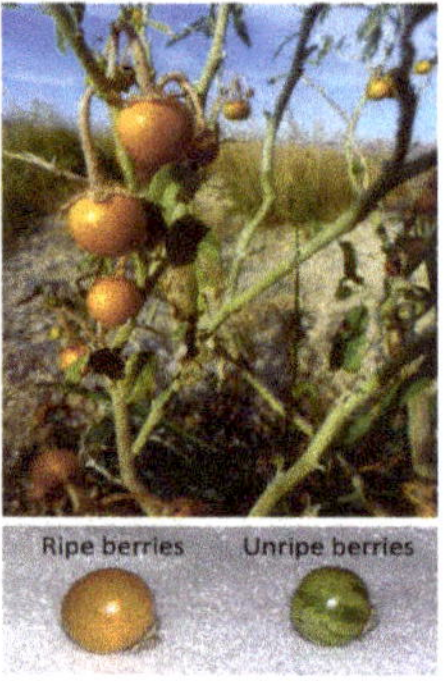

Figure 1. Photograph of the fruits of *Solanum elaeagnifolium*.

2. Materials and Methods

2.1. Plant Material

Ripe yellow fruits were collected from native wild plants of *S. elaeagnifolium* (Figure 1) growing in Chihuahua City and the municipality of Julimes, Mexico. Berries were gathered during the late fall to ensure they were sun-dried. The dry fruits were milled with a mini-mill and sieved with a mesh of 0.5 mm (3383-L19, Thomas Wiley Fisher, Swedesboro, NJ, USA). The powder was stored in hermetically sealed glass vessels at −20 °C. Compositional analysis of the fruits powder was carried out with the AOAC methods 942.05 (ash), 948.22 (fat), 960.52 (protein), 934.01 (moisture) and 991.43 (total, soluble, and insoluble fiber) [15].

2.2. Extraction of Proteases from Fruit Powder

Plant-derived proteases were extracted by mixing the fruit powder with sodium acetate buffer (50 mM, 5%NaCl, pH 5) at a 1:10 ratio (*w*/*v*). The mixture was incubated for 12 h at 4 °C under magnetic stirring and then centrifuged twice at 3200× *g* for 20 min at 4 °C. The supernatant was recovered and filtered first with paper Whatman no. 1 (GE Healthcare, Buckinghamshire, UK) and then through polyethersulfone filters of 0.45 μm (Whatman, GE Healthcare, Buckinghamshire, UK). The content of protein in the plant-derived coagulant was quantified by the Bradford method [16]. The filtered extracts were stored at −20 °C in hermetically sealed plastic tubes.

2.3. Proteolytic Properties of the Plant-Derived Coagulant

The effect that the plant-derived coagulant had on caseins (α_{S1}, α_{S2}, β, and κ) and whey proteins (β-lg, and α-la) were assessed by polyacrylamide gel electrophoresis in the presence of sodium dodecyl sulfate (SDS-PAGE). For this analysis, a solution of casein and a solution of whey proteins were prepared at 1% (*w*/*v*). The reaction of proteolysis was carried out by mixing 130 μL of the protein solution with 25 μL of plant coagulant. After 10 min of incubation at 37 °C, the reaction was stopped by boiling the solution for 15 min. The mixture was cooled to room temperature and loaded in the polyacrylamide gel. The conditions of the SDS-PAGE were according to the methodology of Chavez-Garay et al. [11]. The photographs of gels were digitized and analyzed with the software ImageJ (National Institutes of Health, Bethesda, MD, USA).

The proteolytic activity of the plant-derived coagulant and chymosin was measured in triplicate with a modified Lowry method [17]. Briefly, 130 μL of casein solution (0.66% *w*/*v*) were added with 25 μL of plant coagulant and incubated at 37 °C for 10 min. The proteolytic reaction was ended by adding 130 μL of trichloroacetic acid (100 mM). The hydrolyzed solution of casein was centrifuged at 3000× *g* for 15 min at 10 °C to sediment the non-hydrolized caseins. A volume of 250 μL was taken from the supernatant and placed into an empty tube. To this tube was added 625 μL of Na_2CO_3 (500 mM) and 125 μL of Folin-Ciocalteu solution (0.2 N), and it was incubated at 37 °C for 30 min. The supernatant was read at 760 nm in a microplate reader (Biotek, Elx808, Winooski, VT, USA). Data were interpreted with the calibration curve made with standards of tyrosine.

2.4. Capacity of the Plant-Derived Coagulant to Clot Milk

Milk clotting activity was estimated in both the plant coagulant and chymosin (Chy-Max, Chr Hansen A/S, Horsholm, Denmark) according to the methodology described by Chávez-Garay et al. [11]. Results were expressed as Soxhlet units of coagulation (U) using Equation (1). Soxhlet units or milk clotting units (U) represents the volume of milk that can be clotted by one volume unit of coagulant in 40 min at 35 °C [8]. Thus, one volume of chymosin can clot a larger amount of milk than one volume of the plant-derived coagulant. In this equation, $t_{clotting}$ was the timespan required to clot milk, whereas V_{milk} and $V_{coagulant}$ represent the milliliters of milk and coagulant used in the enzymatic reaction. Additionally, the specific milk clotting activity (*specific* $_{MCA}$) was calculated with Equation (2). [protein] is the concentration of protein expressed as mg/mL of plant coagulant, and $mL_{coagulant}$ is the volume of coagulant used. The *specific* $_{MCA}$ was expressed as U/mg of protein coagulant. Milk clotting activities of the plant coagulant and the chymosin were measured five times.

$$U = \frac{V_{milk}}{V_{coagulant}} \times \frac{40\ min}{t_{clotting}} \quad (1)$$

$$specific\ _{MCA} = \frac{U/mL_{coagulnt}}{(protein)} \quad (2)$$

2.5. Manufacture of Cream Cheese

The milk was heated to 80 °C for 10 min, cooled to 38 °C and poured into glass vessels (Pyrex, Corning, Tewksbury, MA, USA) in portions of 500 mL. Four portions of milk were used for each treatment; consequently, each treatment was analyzed four times. To the milk was added 400 μL of $CaCl_2$ (6.6 M) and inoculated (~1% *w*/*v*) with a freeze-dried starter culture of *L. lactis* spp. (Choozit, Danisco, Niebüll, Germany). The plant-derived coagulant was added at three different volumes in the milk portions. The amounts added were 200, 400 and 800 μL, being equivalent to 0.4, 0.8, and 1.6 μg of protein per milliliter of milk. In the case of cheese made with chymosin (control), milk portions has 10 μL of the coagulant added (0.022 μg of protein/mL of milk). Chymosin from *Aspergillus niger* var. Awamori, kindly donated by Chr Hansen (Chy-Max, Horsholm, Denmark) was used. After the addition of coagulants, the vessels were incubated at 36 °C, reaching a pH of 4.8 (after ~7 h). Curds were then transferred into cloth bags and centrifuged at 1700× *g* for 30 min at 10 °C to remove the whey fraction (Thermo IEC, Needham Heights, MA, USA). The cream cheeses were packed in polyethylene bags and stored at 10 °C. The yields of cream cheeses were calculated with the formula described by Chávez-Garay et al. [11]. All cheeses were subjected to proximate analysis using AOAC methods [15].

2.6. Evaluation of Proteolysis in Cream Cheeses

Proteolysis in cream cheeses was evaluated by SDS-PAGE. The protein fractions in the cream cheeses were obtained by mixing 250 mg of cheese with Tris-HCl buffer pH 8 (165 mM Tris, 1 mM EDTA, 70 mM SDS). The aqueous fractions were loaded onto a polyacrylamide gel (stacking 5%, separating 12%) with enough volume to obtain a concentration of ~45 μg of protein. SDS-PAGE conditions and analysis were as described elsewhere in the text.

2.7. Spreadability of Cream Cheese

Spreadability of the cream cheeses was measured using a texture analyzer TA.XTplus (Stable Micro System, Godalming, UK) with a cell load of 50 kg. Before the analysis, a conical base was filled with cream cheese and stored at 10 °C for two hours. Meanwhile, a plastic (perspex) cone with a 45° angle was fitted in the arm of the texturometer. Then, both the cone and the conical base were aligned vertically in the texturometer. The upper cone penetrated the lower base at a speed of 1 mm/s until it reached a distance of 54 mm, which was 2 mm above the bottom of the lower conical base. The force required to obtain the maximum penetration depth of 2 mm was taken to be the spreadability of the cream cheese [1,18].

2.8. Viscoelastic Properties of Cream Cheese

Viscoelastic properties of cream cheeses were measured with a rheometer AR-2000 (TA Instrument, New Castle, DE, USA) and using dynamic small-amplitude oscillatory tests. Cheese samples were placed between an aluminum (0° angle) plate of 40 mm in diameter and the rheometer platform (gap = 2 mm). Before the stress sweep analysis, the linear viscoelastic region was established by strain sweep analysis at different frequencies. Stress sweeps were carried out with the following test parameters: oscillated stress from 1 to 1000 Pa, logarithmic mode, frequency 1.5 Hz, and temperature of 5 °C. The storage modulus (G'), loss modulus (G''), complex viscosity (η^*), and phase angle (δ) were obtained from the linear viscoelastic region observed during stress sweeps [18]. Results were also expressed in terms of the loss tangent ($\tan \delta$), which represents the dimensionless ratio G''/G'.

3. Results and Discussion

3.1. Gross Composition of S. elaeagnifolium Fruits

The composition of dried-yellow berries from *S. elaeagnifolium* was similar to reported by Chávez-Garay et al. [11], excepting for the fat content (Table 1). The source of fat in the powder of dried fruits was likely the seeds, whereas the fruit peels contributed to the high content of insoluble fiber. Proteins in the dried fruits are probably distributed between the seeds and the fruit peels, since milk clotting activity has been detected in both [11]. The amount of protein was ~14 g/100 g of fruit powder (Table 1). Considering the content of protein and the number of fruits produced by a single plant (40 to 60) [19], these solanaceous fruits could be an available source of proteases.

Table 1. Compositional analysis of the dried yellow berries of *Solanum elaeagnifolium* used in this study.

Component	g/100 g of Dried Fruits
Moisture	5.8 ± 0.1
Fat	5.0 ± 0.1
Ash	5.8 ± 0.08
Protein	13.8 ± 0.3
Total fiber	56.8 ± 2
Soluble fiber	2.9 ± 0.9
Insoluble fiber	53.8 ± 1.1
Carbohydrates [1]	12.8

[1] Estimated by difference.

3.2. Proteolytic Properties of the Plant-Derived Coagulant

The plant coagulant had a proteolytic activity of 1.52 ± 0.097 *U-Tyr*/mL and specific proteolytic activity of 0.78 ± 0.045 *U-Tyr*/mg of protein. In comparison, chymosin had a low proteolytic activity (0.29 ± 0.13 *U-Tyr*/mL) and a high specific activity (2.51 ± 0.08 *U-Tyr*/mg of protein). The augmented specific activity of chymosin was due to the high purity of the enzyme. In counterpart, the plant-derived coagulant had a mixture of diverse proteins in addition to proteases. Figure 2 shows the number of proteins in the plant coagulant and their corresponding molecular weights. Protease in fruits of *S. elaeagnifolium* has a molecular weight (MW) of ~58 kDa [11], that is, the first band in the second lane of Figure 2. A milk clotting enzyme with similar MW (66 kDa) has been reported in the fruits of *Solanum dubium*. This protease (Dubium) was identified as a chymotrypsin-like serine protease by assessment with protease inhibitors [20].

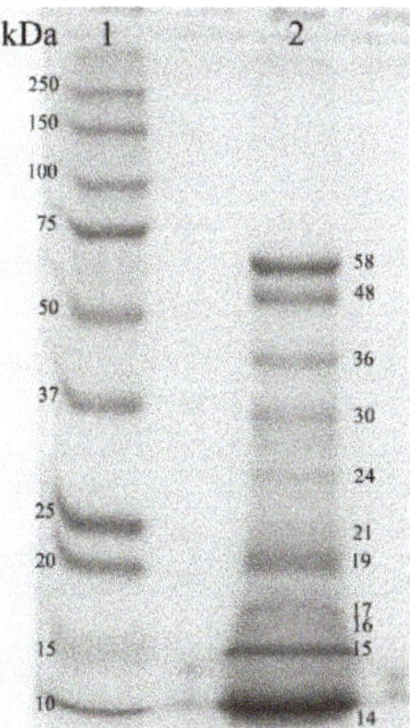

Figure 2. Polyacrylamide gel electrophoresis of the plant coagulant obtained from the yellow berries of *S. elaeagnifolium*. Lane 1: molecular weight marker, lane 2: plant coagulant.

According to the SDS-PAGE analysis, protease from the fruits of *S. elaeagnifolium* was able to hydrolyze caseins (Figure 3a) but not whey proteins (Figure 3b). The plant-derived coagulant hydrolyzed the four types of caseins: α_{s1}-, α_{s2}-, β- and κ-casein. From casein hydrolysis, two peptides (p1 and p2) with MWs of 15 and 14.5 kDa were observed (Figure 3a). However, peptides smaller than 10 kDa (not detected by SDS-PAGE analysis) could have been generated by the plant coagulant. Chymosin hydrolyzed κ-casein and partially α_{S1}-, α_{S2}-, and β-casein (Figure 3a). It is well known that chymosin has a limited proteolytic activity on α_{S1}-, α_{S2}-, and β-casein but is particularly active on κ-casein [21,22]. The specific hydrolysis of chymosin on κ-casein (at Phe_{105}-Met_{106}) releases a hydrophilic glycoprotein (glycomacropeptide) and leaves a positively charged para-κ-casein attached to casein micelles. Consequently, the repulsive electric force among casein micelles decreases, as well as the steric repulsion which favors the clot of milk [22]. On the other hand, the plant-derived coagulant did not hydrolyze the whey proteins (β-lg and α-la) or produce a slight hydrolysis, which could not be detected by SDS-PAGE analysis. In contrast, it was confirmed in the same polyacrylamide gel (Figure 3b) that chymosin hydrolyzed α-la, as reported by other authors [22]. Unlike caseins, whey proteins are globular proteins with a relatively high hydrophobicity and compactly folded peptide chains [23]. Therefore, whey proteins are less susceptible to hydrolysis by proteases than caseins.

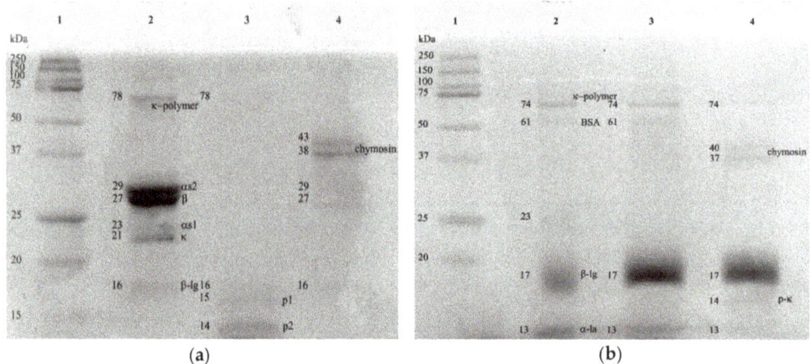

Figure 3. Casein (**a**) and whey protein (**b**) solutions at 1% before (lane 2 in both images) and after (lane 3 in both images) being treated with either the plant coagulant or chymosin (lane 4 in both images) for 10 min at 37 °C. Lane 1 (both images): molecular weight marker. P1 and P2 peptides derived from the hydrolysis of caseins with the plant coagulant.

3.3. Milk Clotting Properties of the Plant-Derived Coagulant

S. elaeagnifolium fruit showed the ability to clot milk (978.36 ± 88.3 *U*/mL), yet not as good as chymosin from *Aspergillus niger* (2285.70 ± 102.3 *U*/mL). The milk clotting activity observed in the plant coagulant was similar to that reported for the fruits of *S. dubium* (880 *U*/mL), a closely related plant [10]. When the specific-milk clotting activities (specific $_{MCA}$) were calculated, the disparity between chymosin and plant coagulant was even more evident (20,779.09 and 923.85 *U*/mg of protein). This increased difference was anticipated, since the purity and quantity of enzyme varied between both coagulants. For instance, 1 mL of chymosin solution contained 0.113 mg of protein or enzyme. In contrast, 1 mL of the plant-derived coagulant had 1.0597 mg of protein, but only a minor fraction of this protein corresponded to proteinases. The ratios of milk clotting activity to proteolytic activity (MCA/PA) were 8278.5 and 1184.4 *U*/*U-Tyr* for chymosin and the plant-derived coagulant, respectively.

Different authors have stated that proteases from plants have a lower capacity to clot milk than rennin or chymosin [9]. The specific hydrolysis of chymosin on κ-casein decreases the zeta potential of casein micelles (~20 mV) by half and reduces the steric repulsion among micelles. These changes greatly favor the aggregation of casein micelles and thus the clot of milk [21,22]. The high milk clotting capacity of chymosin, having a relatively low proteolytic activity, gives this enzyme an elevated MCA/PA ratio. The higher the MCA/PA ratio the better the use of coagulant for cheese making, because the clotting time is short and the volume of coagulant needed is low [21,24]. Proteases in the plant-derived coagulant hydrolyzed not only the κ-casein but also the α_{S1}-, α_{S2}-, and β-casein (as described elsewhere in the text). Even though the hydrolysis of caseins by the plant coagulant was unspecific, it was enough to induce clotting of milk. The MCA/PA ratio of the plant coagulant was lower than chymosin but higher than reported for other plant coagulants like Neriifolin S, papain, trypsin, ficin and religiosin (433, 367, 3.6, 393, and 387 *U*/OD 660 nm) [24,25]. Therefore, proteases from the berries of *S. elaeagnifolium* are suitable for cheese-making. Indeed, this plant coagulant has been used for decades in the manufacture of artisanal asadero cheese in the northwest of Mexico [26].

3.4. Composition of Cream Cheese Made with the Plant-Derived Coagulant

Excepting fat content, the composition of cream cheeses was similar to that reported for commercial brands of either Neufchatel or full-fat cream cheese [1]. The fat content in our cream cheeses ranged from 14–15% (Table 2), whereas commercial Neufchatel and full-fat cream cheese have 20–25% and 32–35% of fat, respectively [1]. As previously described, cream cheeses were manufactured using three different concentrations of plant coagulant (0.4, 0.8, and 1.6 µg protein/mL milk) and the control cheeses were made with chymosin. One primary concern for using plant proteases for cheese-making is the loss of fat and proteins in the cheese whey, which reduces the cheese yield. This problem arises from the excessive and unspecific proteolysis that some plant coagulants have on caseins, leading to the formation of weak milk-gel structures [21]. Proteases from the fruits of *S. elaeagnifolium* modified neither fat, lactose, nor ash content in cream cheeses at any plant coagulant concentration (Table 2). The maximal concentration of plant-derived coagulant brought about a cream cheese slightly wetter, with lower protein content than control cheese (Table 2). However, the yields of cream cheeses manufactured with the plant-derived coagulant were only marginally reduced (statistically not significant), regardless of the concentration of plant coagulant used (Table 2).

Table 2. Composition and yield of cream cheeses made with chymosin and a plant coagulant obtained from the ripe fruits of *S. elaeagnifolium*.

(PROTEASE Added) µg of Protein/mL of Milk	(0.022)	(0.4)	(0.8)	(1.6)
Component	Control [1]	Cheese made with plant coagulant		
Moisture (g/100 g)	59.7 ± 8.8^{a}	59.5 ± 6.5^{a}	58.5 ± 4.4^{a}	61.9 ± 2.8^{a}
Fat (g/100 g)	14.4 ± 0.6^{a}	14.4 ± 0.7^{a}	14.9 ± 0.1^{a}	14.3 ± 0.7^{a}
Protein (g/100 g)	6.0 ± 0.4^{a}	5.1 ± 0.2^{ab}	5.3 ± 0.8^{ab}	4.7 ± 0.6^{b}
Ash (g/100 g)	1.1 ± 0.1^{a}	1.0 ± 0.05^{a}	1.0 ± 0.1^{a}	0.8 ± 0.2^{a}
Carbohydrates [2] (g/100 g)	18.8 ± 2.0^{a}	20.1 ± 1.8^{a}	20.3 ± 1.9^{a}	18.3 ± 2.1^{a}
Yield (%)	16.1 ± 1.21^{a}	14.9 ± 0.74^{a}	14.0 ± 2.4^{a}	15.0 ± 2.1^{a}

[1] Cheese made with chymosin from *Aspergillus niger* var. *awamori* (Chr Hansen). [2] Calculated by difference. SE = standard error. [a, b] Means followed by different letters in the same row are statistically different (ANOVA, Tukey-Kramer test, $p < 0.05$). Data represent the average of four replicates (*n*).

3.5. Proteolysis in Cream Cheeses

The gel electrophoresis analysis unveiled that all caseins (αs_1-, αs_2-, β-, and κ-casein) were partially hydrolyzed in cream cheeses manufactured with the plant-derived coagulant. Conversely, chymosin extensively degraded κ-casein in cream cheeses but scarcely hydrolyzed other type of caseins (Figures 4 and 5). The incremental addition of plant-derived coagulant enhanced the proteolysis of caseins in cream cheeses, particularly the breakdown of αs_1-casein (Figure 5). SDS-PAGE analysis of the cheeses also confirmed the low proteolytic activity that plant protease has on caseins. As previously described, the plant-derived coagulant had lower specific proteolytic activity on caseins (0.78 ± 0.045 *U-Tyr*/mg of protein) than chymosin (2.51 ± 0.08 *U-Tyr*/mg of protein). It is worthwhile highlighting these results, because most plant proteases exhibit high levels of proteolytic activity on caseins, which leads to the formation of multiple small peptides. Formation of such peptides can impair texture and produce a bitter flavor in the cheese [27,28]. Only four peptides were detected in the cheese made with the plant coagulant: p1 = 15 kDa, p2 =14.5 kDa, p3 = 12.8 kDa, p4 = 12.5 kDa (Figure 4).

Bitter peptides are mostly hydrophobic and arise from the breakdown of α_{s1}-, and β-casein [29]. Some authors have suggested that bitter peptides produced by milk coagulants (such as calf rennet, microbial, and plant proteases like cardosin) have a MW of 0.15 to 14 kDa [29]. Other authors have reported that bitter peptides consist of 2 to 23 amino acid residues with MW ranging from 0.5 to 3 kDa [30].

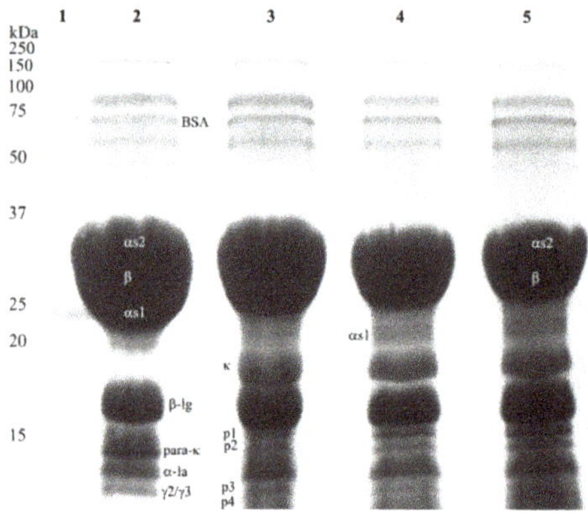

Figure 4. SDS-PAGE of aqueous extracts obtained from cream cheeses. Lane 1: molecular weight marker; lane 2: cheese made with chymosin; lane 3: cheese clotted with a plant coagulant (0.042 mg of protein/100 mL of milk); lane 4: cheese clotted with a plant coagulant (0.0846 mg of protein/100 mL of milk); lane 5: cheese clotted with a plant coagulant (0.1694 mg of protein/100 mL of milk).

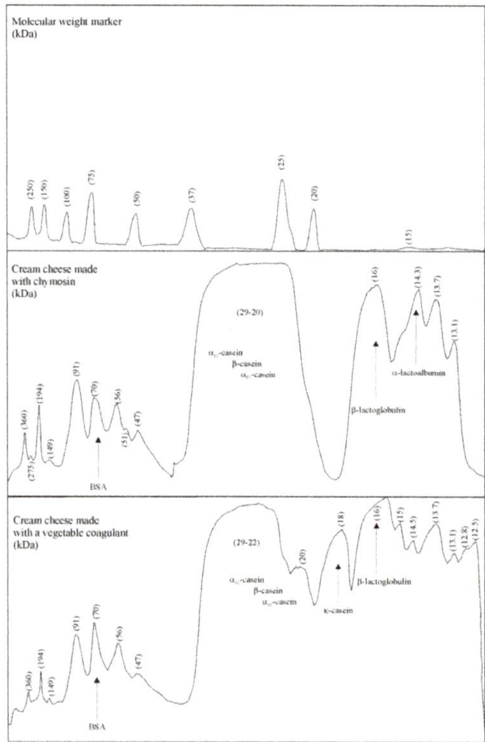

Figure 5. Densitometric analysis of SDS-PAGE with samples of cream cheese made with chymosin and cream cheese made with the plant coagulant obtained from the berries of *S. elaeagnifolium*.

3.6. Rheological Properties of Cream Cheese

Cream cheeses clotted with the fruits of *S. elaeagnifolium* had spreadability values comparable to those coagulated with chymosin (the control cheese). However, the viscoelastic properties of the cream cheeses varied significantly depending on the coagulant and volume of coagulant used (Table 3). For instance, cream cheese clotted with a small quantity of plant coagulant (0.4 mL of coagulant/L milk) had lower values of η^*, G', and G'' than control chymosin cheese. Doubling the volume of plant coagulant (0.8 mL of coagulant/L milk) resulted in cream cheese with viscoelastic properties similar to the control cheese. However, the opposite effect was observed with a further volume increase of plant-derived coagulant (1.6 mL of coagulant/L milk), which notably decreased the viscoelastic properties of cream cheese.

Table 3. Rheological properties of cream cheeses made with chymosin and a plant coagulant obtained from the ripe berries of *S. elaeagnifolium*.

(Protease Added) µg of Protein/mL of Milk	(0.022)	(0.4)	(0.8)	(1.6)
Item	Control [1]	Cheese made with plant coagulant		
Spreadability [2] (N)	151.3 ± 27.8 [a]	117.7 ± 8.1 [a]	151.6 ± 23.3 [a]	101.3 ± 25.8 [a]
η^* (kPa×s)	20.4 ± 7.8 [ab]	10.5± 3.9 [bc]	29.2 ± 11.2[a]	4.4 ± 0.77 [c]
G' (kPa)	124.3 ± 28.9 [b]	64.1 ± 35.5 [bc]	178.6 ± 18.9 [a]	26.8 ± 4.8 [c]
G'' (kPa)	32.1± 13.9 [a]	15.8 ± 4.4 [b]	42.8 ± 13.3 [a]	7.4 ± 2.9 [b]

[1] Cheese made with chymosin from *Aspergillus niger* var. *awamori* (Chr Hansen). [2] Force required to spread the cheese between two surfaces. SE = standard error. η^* = complex viscosity. G' = storage modulus. G'' = loss modulus. [a, b] Means followed by different letters in the same row are statistically different (ANOVA, Tukey-Kramer test, $p < 0.05$). Data represent the average of four replicates (n).

Addition of rennet during gel formation in acid-curd cheeses (like cream cheese) accelerate milk clotting and induce a coarser protein network that enhances the gel firmness [3,7]. It is, however, necessary to add an adequate amount of coagulant in the milk (critical rennet concentration, CRC) to synchronize the acid and rennet clotting. Otherwise, a mild or adverse effect could be observed by a low or excessive addition of rennet [3]. It was noted that cream cheeses with 0.8 mL of plant coagulant per liter of milk added had the lowest value of tan δ—0.239 ± 0.01 (or highest G' value). It is likely that such plant coagulant concentration not only hydrolyzed caseins but also did so at the proper rate. Thus, cross-linking between proteins (caseins and whey proteins) and peptides could have been favored. Nevertheless, a lower concentration of plant coagulant (0.4 mL of coagulant/L milk) produced insufficient hydrolysis of caseins (Figure 4) which increased the value of tan δ (0.246 ± 0.05). Similarly, a concentration of plant coagulant higher than 0.8 mL of coagulant/L milk led to extensive hydrolysis of caseins (Figure 4), raising tan δ to 0.276 ± 0.05. The increment of tan δ in a combined rennet and bacterial fermentation indicates an imbalance of negative charges on the micelles, which weakens the milk-gel structure [7,31].

Cream cheeses made with chymosin had a tan δ value (0.257 ± 0.015) higher than that observed in cheeses made with 0.8 mL of plant-derived coagulant per liter (tan δ = 0.239 ± 0.01). As previously described, the plant protease degraded all the caseins (α_{S1}-, α_{S2}-, β-, and κ-casein) slightly, releasing four peptides (p1, p2, p3, and p4). In contrast, chymosin extensively broke up κ-casein with a concomitant formation of para-κ-casein and a glycomacropeptide (Figures 3 and 5). It is possible that peptides released by the plant-derived coagulant enhanced the interactions among caseins in a better way than casein fractions released by chymosin.

4. Conclusions

S. elaeagnifolium is considered a weed by farmers. However, the dry fruits of this plant contain a milk clotting protease at high concentration. This plant-derived protease had a lower ratio of milk clotting activity to proteolytic activity (*MCA/PA*) than chymosin, but a *MCA/PA* higher than reported for other plant proteases. With reference to cream cheese, its yields and composition did not change with the use of the plant-derived coagulant. Only the maximal concentration of plant-derived coagulant brought about a cream cheese slightly wetter and with lower protein content than a control cheese. SDS-PAGE analysis revealed that proteases from *S. elaeagnifolium* hydrolyzed all caseins, but only partially. From this hydrolysis, four peptides ranging from 12.5 to 15 kDa were observed. The unspecific hydrolysis produced by the plant protease modified the viscoelastic properties of the cream cheeses produced. We found that the critical enzyme concentration (CRC) for this plant coagulant is 0.8 mL of plant coagulant per liter of milk (0.8 μg of total protein per mm of milk). According to the results obtained, this concentration enhances the interaction between proteins and peptides, leading to an increase in the viscoelastic properties of cream cheese.

Author Contributions: A.B.-C., S.E.G.-S. and P.R.-V. performed the experimental trials; M.Y.L.-R. participated in the interpretation of rheological assays. A.G.-T. analyzed the results of proteolysis. Meanwhile, N.G.-M. conceived the study and drafted the manuscript.

Funding: This research received no external funding.

Conflicts of Interest: The authors declare no conflict of interest.

References

1. Brighenti, M.; Govindasamy-Lucey, S.; Lim, K.; Nelson, K.; Lucey, J.A. Characterization of the rheological, textural, and sensory properties of samples of commercial US cream cheese with different fat contents. *J. Dairy Sci.* **2008**, *91*, 4501–4517. [CrossRef] [PubMed]
2. Coutouly, A.; Riaublanc, A.; Axelos, M.; Gaucher, I. Effect of heat treatment, final pH of acidification, and homogenization pressure on the texture properties of cream cheese. *Dairy Sci. Technol.* **2013**, *94*, 125–144. [CrossRef]
3. Schulz-Collins, D.; Senge, B. Acid-and acid/rennet-curd cheeses part A: Quark, cream cheese and related varieties. In *Cheese: Chemistry, Physics and Microbiology*; Elsevier: New York, NY, USA, 2004; pp. 301–328.
4. Lucey, J.A.; Singh, H. Formation and physical properties of acid milk gels: A review. *Food Res. Int.* **1997**, *30*, 529–542. [CrossRef]
5. Cosentino, C.; Paolino, R.; Freschi, P.; Calluso, A.M. Short communication: Jenny milk as an inhibitor of late blowing in cheese: A preliminary report. *J. Dairy Sci.* **2013**, *96*, 3547–3550. [CrossRef] [PubMed]
6. Lucey, J.; Singh, H. Acid coagulation of milk. In *Advanced Dairy Chemistry—1 Proteins*; Fox, P.F., McSweeney, P.L.H., Eds.; Springer: New York, NY, USA, 2003; pp. 1001–1025.
7. Salvatore, E.; Pirisi, A.; Corredig, M. Gelation properties of casein micelles during combined renneting and bacterial fermentation: Effect of concentration by ultrafiltration. *Int. Dairy J.* **2011**, *21*, 848–856. [CrossRef]
8. Jacob, M.; Jaros, D.; Rohm, H. Recent advances in milk clotting enzymes. *Int. J. Dairy Technol.* **2011**, *64*, 14–33. [CrossRef]
9. Shah, M.A.; Mir, S.A.; Paray, M.A. Plant proteases as milk clotting enzymes in cheese-making: A review. *Dairy Sci. Technol.* **2014**, *94*, 5–16. [CrossRef]
10. Ahmed, I.A.M.; Morishina, I.; Babiker, E.E.; Mori, N.; Morishima, I.; Babiker, E.E.; Mori, N. Characterisation of partially purified milk clotting enzyme from Solanum dubium Fresen seeds. *Food Chem.* **2009**, *116*, 395–400. [CrossRef]
11. Chávez-Garay, D.R.; Gutiérrez-Méndez, N.; Valenzuela-Soto, M.E.; García-Triana, A. Partial characterization of a plant coagulant obtained from the berries of Solanum elaeagnifolium. *CyTA J. Food* **2016**, *14*, 200–205. [CrossRef]
12. Gutiérrez-Méndez, N.; Chávez-Garay, D.R.; Jiménez-Campos, H. Exploring the milk clotting properties of a plant coagulant from the berries of S. elaeagnifolium var. Cavanilles. *J. Food Sci.* **2012**, *71*, 89–94. [CrossRef]
13. Bodansky, A. The chymase of Solanum elaeagnifolium a preliminary note. *J. Biol. Chem.* **1916**, *27*, 103–105.
14. Bodansky, A. A study of a milk-coagulating enzyme of Solanum elaeagnifolium. *J. Biol. Chem.* **1924**, *61*, 365–375.
15. AOAC International. *Official Methods of Analysis of AOAC International*, 16th ed.; Cunnif, P., Ed.; Association of Official Analytical Chemists International: Arlington, MA, USA, 1998.

16. Bradford, M.M. A rapid and sensitive method for the quantitation of microgram quantities of protein utilizing the principle of protein-dye binding. *Anal. Biochem.* **1976**, *72*, 248–254. [CrossRef]
17. McDonald, C.E.; Chen, L.L. The Lowry modification of the Folin reagent for determination of proteinase activity. *Anal. Biochem.* **1965**, *10*, 175–177. [CrossRef]
18. Almanza-Rubio, J.L.; Gutiérrez-Méndez, N.; Leal-Ramos, M.Y.; Sepulveda, D.; Salmeron, I. Modification of the textural and rheological properties of cream cheese using thermosonicated milk. *J. Food Eng.* **2016**, *168*, 223–230. [CrossRef]
19. OEPP. Solanum elaeagnifolium. *Bull. OEPP* **2007**, *37*, 236–245. [CrossRef]
20. Ahmed, I.A.M.; Morishima, I.; Babiker, E.E.; Mori, N. Dubiumin, a chymotrypsin-like serine protease from the seeds of Solanum dubium Fresen. *Phytochemistry* **2009**, *70*, 483–491. [CrossRef]
21. Horne, D.S.; Banks, J.M. Rennet-induced coagulation of milk. In *Cheese Chemistry, Physics and Microbiology*; Fox, P.F., McSweeney, P.L.H., Cogan, T.M., Guinee, T.P., Eds.; Elsevier Academic Press: San Diego, CA, USA, 2004; pp. 47–70.
22. Crabbe, M. Rennets: General and molecular aspects. In *Cheese: Chemistry, Physics and Microbiology*; Elsevier Academic Press: San Diego, CA, USA, 2004; pp. 19–45.
23. Walstra, P.; Wouters, J.T.M.; Geurst, T.M. *Dairy Science and Technology*, 2nd ed.; CRC Press: Boca Raton, FL, USA, 2006.
24. Yadav, R.P.; Patel, A.K.; Jagannadham, M.V. Neriifolin S, A dimeric serine protease from Euphorbia Neriifolia Linn.: Purification and biochemical characterisation. *Food Chem.* **2012**, *132*, 1296–1304. [CrossRef]
25. Kumari, M.; Sharma, A.; Jagannadham, M.V. Decolorization of crude latex by activated charcoal, purification and physico-chemical characterization of religiosin, a milk clotting serine protease from the latex of *Ficus religiosa*. *J. Agric. Food Chem.* **2010**, *58*, 8027–8034. [CrossRef]
26. Martínez-Ruiz, N.R.; López-Díaz, J.A. Optimización de la extracción y estandarización de un cuajo vegetal para la elaboración de queso asadero. *Cienc. Front.* **2008**, *6*, 173–176.
27. Egito, A.S.; Girardet, J.M.; Laguna, L.E.; Poirson, C.; Mollé, D.; Miclo, L.; Humbert, G.; Gaillard, J.L. Milk clotting activity of enzyme extracts from sunflower and albizia seeds and specific hydrolysis of bovine κ-casein. *Int. Dairy J.* **2007**, *17*, 816–825. [CrossRef]
28. Puglisi, I.; Petrone, G.; Lo Piero, A.R. A kiwi juice aqueous solution as coagulant of bovine milk and its potential in Mozzarella cheese manufacture. *Food Bioprod. Process.* **2014**, *92*, 67–72. [CrossRef]
29. Agboola, S.; Chen, S.; Zhao, J. Formation of bitter peptides during ripening of ovine milk cheese made with different coagulants. *Dairy Sci. Technol.* **2004**, *84*, 567–578. [CrossRef]
30. Karametsi, K.; Kokkinidou, S.; Ronningen, I.; Peterson, D.G. Identification of bitter peptides in aged cheddar cheese. *J. Agric. Food Chem.* **2014**, *62*, 8034–8041. [CrossRef]
31. Cosentino, C.; Paolino, R.; Musto, M.; Freschi, P. Innovative use of jenny milk from sustainable rearing. In *The Sustainability of Agro-Food and Natural Resource Systems in the Mediterranean Basin*; Vastola, A., Ed.; Springer International Publishing: Cham, Switzerland, 2015.

 *foods*

Article

Salt Distribution in Raw Sheep Milk Cheese during Ripening and the Effect on Proteolysis and Lipolysis

Olaia Estrada [1], Agustín Ariño [2,*] and Teresa Juan [1,2]

[1] Centro de Investigación y Tecnología Agroalimentaria de Aragón (CITA), 50059 Zaragoza, Spain; oestrada@cita-aragon.es (O.E.); tjuan@cita-aragon.es (T.J.)

[2] Instituto Agroalimentario de Aragón-IA2 (Universidad de Zaragoza-CITA), Facultad de Veterinaria, 50013 Zaragoza, Spain

* Correspondence: aarino@unizar.es; Tel.: +34-876-554142

Received: 7 February 2019; Accepted: 12 March 2019; Published: 17 March 2019

Abstract: The salt distribution process in artisanal sheep cheese with an innovative shape of eight lobes was investigated. The cheese was subjected to two brining conditions: 24 h with brine at 16°Baumé and 12 h at 22°Baumé. The chemical composition (pH, water activity, dry matter, fat, and protein content), proteolysis (nitrogen fractions and free amino acids), and lipolysis (free fatty acids) were evaluated in two sampling zones (internal and external) at 1, 15, 30, 60, 90, 120, 180, and 240 days of ripening. The whole cheese reached a homogeneous salt distribution at 180 days of ripening. Brining conditions did not have an influence on the rate of salt penetration, but on the final sodium chloride (NaCl) content. Cheese with higher salt content (3.0%) showed increased proteolysis and lipolysis as compared to cheese with lower salt content (2.2%). Proteolysis index and total free fatty acids did not differ significantly ($p > 0.05$) between internal and external zones of cheese. It is suggested that producers start marketing this artisanal cheese at 6 months of ripening, when it has uniform composition and salt distribution.

Keywords: raw sheep milk cheese; artisanal cheese; salt distribution; proteolysis; lipolysis; ripening; sampling zone

---

1. Introduction

The consumer interest in and demand for artisanal dairy products are increasing. The development of these products elaborated on a small scale is receiving more consideration from European policy-makers due to their important role in the local agricultural economy and their contribution to development and sustainability, especially in rural areas.

Cheeses from Teruel (Spain) have been acclaimed since time immemorial. Miguel de Cervantes mentioned Tronchon cheese from Teruel in one of the greatest works of literature, *The Ingenious Nobleman Sir Quixote of La Mancha*, in 1615. Teruel cheese is a hard cheese produced with raw sheep or goat milk (not mixed), maintaining the traditional manufacturing process but including an innovative mold of eight lobes, unique in the market. This new shape, joining tradition and innovation, was designed and patented by producers to attract the attention of consumers.

In artisanal cheese varieties, flavor is more intense and rich than in industrially processed cheeses due to the native microbiota that develop peculiar organoleptic characteristics, which are appreciated by consumers [1]. Salt content plays a key role in the complex biochemical reactions during cheese ripening due to its effect on microbial growth, enzyme activity, and syneresis [2]. The main methods of salting are adding sodium chloride to the curd (dry salting), rubbing the salt on the surface of the cheese (dry surface rubbing), or immersing the cheese in a brine solution of sodium chloride and water (brining). In Spain, most semi-hard and hard cheeses are salted by brining.

During brining and ripening, salt diffusion and outward migration of water take place until equilibrium is reached throughout the cheese. Although this is a crucial step in cheese manufacture, there may be considerable variability in brining conditions, such as time, salinity, and temperature, for the same cheese variety [3]. In spite of the effect of salt content on organoleptic characteristics of cheese, only five out of 28 Spanish cheeses with Protected Designation of Origin (PDO) status have established the sodium chloride content in their respective regulations. Regulations for PDO Manchego (<2.3%), PDO Torta del Casar (<3%), PDO Valdeon (<3.5%), and PDO L'Alt Urgell y La Cerdanya (1.5–2.5%) cheeses detail sodium chloride criteria.

Proteolysis and lipolysis are the most complex biochemical changes that occur during cheese ripening. A reduced salt content in cheese has been related to increased proteolysis and sensory defects such as reduction of firmness, excessive acidity, and bitter taste [4,5]. Lipolysis seems to have less correlation with the salt content of cheese; free fatty acids released during ripening were not modified by salt reduction in Cheddar cheese [6]. Only a few studies have investigated the effects of salt distribution on proteolysis and lipolysis during cheese ripening. The sampling zone (internal and external) has been considered in large-format cow's milk cheeses such as Parmigiano-Reggiano (>30 kg) [7,8], Reggianito Argentino (5–10 kg weight) [9], and Port Salut Argentino [10]. However, there are no studies available on the effect of salt distribution on the phenomena of proteolysis and lipolysis during ripening in raw sheep milk cheese. Additionally, the shape of cheese may affect the rate of salt absorption [2], so it is important to elucidate details of how the salt is distributed in Teruel cheese, with its innovative multilobed shape.

The aim of the current study was to assess the salt distribution process in a hard cheese (~4 kg) made with raw sheep milk and ripened for 240 days, elaborated under artisanal cheesemaking methods and with a multilobed shape. Two brining conditions were tested (24 h at 16°Baumé versus 12 h at 22°Baumé). In addition, the effect of salt content on chemical composition, proteolysis, and lipolysis was evaluated at two sampling zones (internal and external). The results obtained are intended to provide producers with important scientific information on the optimal time to put the cheese on the market.

2. Materials and Methods

2.1. Cheese-Making Procedure

In order to select the conditions for the study, we visited 10 artisanal dairies belonging to the Asociación Turolense de Productores de Leche y Queso (Association of Milk and Cheese Producers from Teruel). The cheesemaking process was analyzed and showed many similarities among dairies; most of the differences were due to the salting process. Thus, two dairies representing the two most common salting procedures were selected to carry out the experiments, in which the main objective was to study the degree of uniformity in the rate of absorption of salt by both brining conditions, considering the new shape with eight lobes. The operating conditions at both dairies were identical in relation to the type and amount of rennet used, the type and inoculation dosage of starter culture, the duration and temperature of coagulation, curd cutting and agitation, grain size of the curd, and molding and pressing. The only difference between dairies was the brining conditions, as described below. The cheeses were manufactured according to traditional methods (Figure 1) by two farmhouse cheese producers located in Teruel, Spain (named dairy plants D1 and D2), with raw sheep milk from Assaf breed produced by their own livestock. The chemical composition of the milk (g/100 g) was fat 6.79 $\pm$ 1.51, protein 5.30 $\pm$ 0.52, lactose 4.81 $\pm$ 0.66, and nonfat solids 11.03 $\pm$ 0.30. Milk was stored at 4 $\pm$ 1 °C until used (48 h maximum).

Figure 1. Cheesemaking procedure of artisanal raw sheep milk cheese: (**A**) Assaf sheep; (**B**) milk cooling tank; (**C**) cheese vat; (**D**) curd cutter; (**E**) manual cutting; (**F**) draining the whey; (**G**) dry curd; (**H**) manual molding; (**I**) pressing; (**J**) brine tank; (**K**) brining; (**L**) maturation chamber; (**M**) one-day ripened Teruel cheese; (**N**) three months of ripening; (**O**) six months of ripening.

Lyophilized heterofermentative culture (CHOOZIT 4001/4002 MA, Danisco) was dispersed in approximately 300 mL of fresh milk and then added to the manufacturing milk (2%, w/v) at 20 °C. Starter cultures contained *Lactococcus lactis* subsp. *lactis*, *Lactococcus lactis* subsp. *cremoris*, *Lactococcus lactis* subsp. *lactis* biovar diacetylactis, and *Streptococcus thermophilus*. Milk was heated to 30 ± 1 °C and bovine rennet (~70% chymosin, ~30% pepsine; Arroyo, Spain) was added at a ratio of 25 mL/100 L to obtain clotting within 30–35 min. When the master cheesemaker observed the proper consistency of the curd, the coagulum was cut into rice/chickpea-sized grains (5–10 mm) and stirred for 45 min at 35 °C. Then, the whey was drained off and the curd was manually placed in microperforated plastic molds with an 8-lobed shape. Cheeses were pressed in a horizontal pneumatic press increasing from 2 to 3.5 kPa for nearly 6 h and then salted by immersion in brine (NaCl solution) at 10 °C. At dairy D1, the brine strength was 16°Baumé for 24 h, and at D2 the brine strength was 22°Baumé for 12 h. After 2–3 days in the airing area, the cheeses were stored in a maturation chamber at 10–12 °C and 80–85% relative humidity and were turned over periodically during ripening. The cheeses had a shape of eight lobes, a diameter of 27 cm, a height of 15 cm, and a weight 4–5 kg, depending on the ripening stage.

2.2. Sampling

Three independent replicate trials of cheeses were produced within an interval of 15 days at each dairy plant (D1 and D2), and eight cheeses were manufactured for each batch, for a total of 48 cheeses. For the chemical analysis, one cheese from each batch was taken randomly at 1, 15, 30, 60, 90, 120, 180, and 240 days of ripening. Cheeses were transported to the laboratory under refrigerated conditions (below 4 °C) and two sampling zones were studied, the external zone (Ext) and the internal zone (Int). A vertical cylinder section (6 cm diameter) cut from the cheese center was considered the internal zone (Int). After removing the rind (1 cm thick), eight cylinders of 3 cm in diameter (one per lobe) were cut perpendicular to the flat surface to sample the external zone (Ext). Cheese samples were analyzed the same day they were received for pH, dry matter content, and water activity. Aliquots of each sample were vacuum-packed and kept at −20 °C until the rest of the analytical determinations. A total of 96 samples (2 dairy plants × 3 batches × 8 ripening times × 2 sampling zones) were analyzed and each assay was carried out in duplicate.

2.3. Physicochemical Analysis

Water activity (a_W) was measured directly using an Aqualab CX-2 dew-point hygrometer (Decagon Devices Inc., Pullman, USA). For pH measurement, ground cheese (10 g) was macerated with 50 mL of distilled water, and the pH of the resultant slurry was measured using an electrode Ag/AgCl (Crison pH 52–21) at 20 °C. Dry matter (DM) was analyzed according to the standards of the International Dairy Federation (ISO 5534:2004/IDF 4:2004) [11]. Total fat content was determined by the gravimetric method (ISO1735:2004/IDF 005) [12] in a Soxtec Avanti 2055 (Foss Tecator, Höganäs, Sweden) and drying in an oven (Memmert, Schwabach, Germany). Total nitrogen (TN) was determined by the Kjeldahl method (ISO 8961-1) [13] using a Kjeltec 8400 Analyzer (Foss Analytical, Höganäs, Sweden). Total protein content was calculated as TN multiplied by 6.38. Salt content (NaCl) was determined using an automatic titrator 848 Titrino Plus (Metrohm, Herisau, Switzerland) according to ISO 5943:2006/IDF 088 [14]. The method is based on potentiometric titration of chloride ions with a solution of silver nitrate (0.1 mol/L) and using the conversion factor 58.4 to calculate the concentration of sodium chloride in g NaCl/100 g DM. All analyses were performed in duplicate.

2.4. Proteolysis Assessment

The nitrogen fraction water soluble nitrogen (WSN), pH 4.4 soluble nitrogen (pH 4.4-SN), trichloroacetic acid 12% (*w*/*v*) soluble nitrogen (TCA-SN), and phosphotungstic acid soluble nitrogen (PTA-SN) were determined by International Dairy Federation FIL-IDF 25:1964 according to Butikofer et al. [15]. Four proteolytic indices were calculated as ratios between the various soluble fractions related to total nitrogen (TN): WSN/TN and pH 4.4-SN/TN as the ripening extension index, TCA-SN/TN as the ripening depth index, and PTA-SN/TN as the free amino acid index.

Free amino acids (FAAs) were determined with AccQ.Fluor Reagent® from Waters. The free amino acid extraction was performed according to Milesi et al. [16] with some modifications. Cheese samples (5 g) were homogenized with 15 mL of distilled water and norleucine (500 μL, 25 mM) as internal standard using a Multi Reax shaker (Heidolph, Schwabach, Germany) for 15 min at 2000 rpm. The suspension was incubated in a water bath at 40 °C for 1 h and centrifuged at 4500 rpm for 30 min at 4 °C. The supernatant was filtered through fast-flow filter paper and the volume was made up to 25 mL with HCl 0.1 N. Afterward, the extract was filtered through 0.45 μm Millex membranes (Millipore, Molsheim, France) and diluted with HCl 0.1 N (1:4). Ten microliters of the diluted solution were mixed with 70 μL of borate buffer 0.2 M (pH 8.8) and 20 μL of the derivatization solution (10 mM; 6-aminoquinolyl-N-hydroxysuccinimidyl carbamate, AQC) as described by the manufacturer [17]. The reaction mixture was kept at room temperature for 1 min and then incubated in an oven at 55 °C for 10 min. The derivatized FAAs were separated, identified, and quantified by a reverse-phase high-performance liquid chromatography (HPLC) system Agilent 1100 Series (Agilent Technologies, Waldbronn, Germany) equipped with a diode-array detector (DAD) and a fluorescence detector (FLD). Chromatographic separation was carried out using a precolumn connected to a Kinitex 2.6 μm C18 100A column (100 mm × 4.6 μm) from Phenomenex (Torrence, CA, USA) at a temperature of 37 °C. Detection was performed with a fluorescence detector using an excitation wavelength of 250 nm and an emission wavelength of 395 nm, except for tryptophan, which was detected by DAD set at 254 nm. The injection volume was 5 μL and the flow rate was 1.3 mL/min. Data acquisition and processing were done with a ChemStation for LC system from Agilent Technologies.

Using the AQC method proposed by Waters Corporation, good resolution was achieved for 17 amino acid standards. However, for a complex matrix such as cheese, more amino acids are present, so a modified AQC derivatization method was needed. In this study, 25 amino acids were evaluated following the method proposed by Estrada et al. [18]: aspartic acid (Asp), asparagine (Asn), serine (Ser), glutamic acid (Glu), glycine (Gly), histidine (Hys), glutamine (Gln), taurine (Tau), arginine (Arg), citrulline (Cit), threonine (Thr), alanine (Ala), proline (Pro), γ-aminobutyric acid (GABA), α-aminobutyric acid (AABA), cysteine (Cys), tyrosine (Tyr), valine (Val), methionine (Met), ornithine

(Orn), lysine (Lys), isoleucine (Ile), leucine (Leu), phenylalanine (Phe), and tryptophan (Trp). With the elution gradient shown in Supplementary Table S1, all AQC derivatives were eluted in 24 min.

2.5. Lipolysis Assessment: Analysis of Free Fatty Acids

Free fatty acids (FFAs) were analyzed according to the method described by Chavarri et al. [19] based on de Jong and Badings's method [20]. Ground cheese (1.0 g) was homogenized with 3.0 g anhydrous sodium sulfate (Na_2SO_4), 0.3 mL 2.5 M sulfuric acid (H_2SO_4), and 100 µL internal standard solution (1 mg/mL pentanoic, nonanoic, and heptadecanoic acids). Lipids were extracted three times with 3 mL diethyl ether-heptane (1:1, v/v) and the solution was clarified by centrifugation. Organic phases containing FFAs and triacylglycerols were combined. The lipid extract was fractionated and triacylglycerols were separated from the FFAs on Sep-Pak aminopropyl Vac cartridges (Waters Corporation, Milford, Massachussetts) using a solid phase extraction (SPE) device GX-271 ASPEC (Gilson, Middleton, WI, USA). The column was equilibrated with 10 mL of heptane. After loading the lipid extract, triacylglycerols were eluted with 10 mL of chloroform–propanol (2:1, v/v) and FFAs were eluted with 5 mL diethyl ether containing 2% formic acid. The extract was analyzed the same day of extraction by gas chromatography (GC) on a HP 6890 Series GC (Agilent Technologies, Waldbronn, Germany) equipped with a flame ionization detector (FID) and HP-Innowax column (30 m × 0.32 mm id × 0.25 µm film thickness). Chromatographic conditions were as follows: 10 min at 65 °C, then up to a final temperature of 240 °C at 10 °C/min. Helium flow was set at 1 mL/min and a split ratio of 1:5. The FFAs were quantified by the internal standard method. Calibration curves were prepared by combining increasing concentrations (from 10 to 800 mg/L) of a mixture of butyric (C4:0), caproic (C6:0), caprylic (C8:0), capric (C10:0), lauric (C12:0), myristic (C14:0), palmitic (C16:0), palmitoleic (C16:1), stearic (C18:0), oleic (C18:1), linoleic (C18:2), and linolenic (C18:3) fatty acids. Ratios between the different fractions were calculated: short-chain fatty acids (SCFAs; C4:0–C8:0), medium-chain fatty acids (MCFAs; C10:0–C14:0), and long-chain fatty acids (LCFAs; C16:0–C18:3).

2.6. Statistical Analysis

Statistical analysis was performed using SPSS 19.0 (IBM Corp., Armonk, NY, USA). Data were analyzed using the general linear model (GLM) procedure and analysis of variance (ANOVA) was performed to establish the presence or absence of significant differences in the chemical composition, proteolysis, and lipolysis considering dairy plant and sampling zone as main factors and ripening time (1, 15, 30, 60, 90, 120, 180, and 240 days) as a co-variable. A p-value < 0.05 was considered statistically significant. Principal component analysis (PCA) was applied to nitrogen fractions and FAA data to reduce the variables to a minimum number of factors. Factors were rotated using the Varimax method to interpret the results.

3. Results and Discussion

3.1. Chemical Composition

The average values of pH, water activity (a_W), dry matter (DM), fat, and protein in external and internal zones of cheeses manufactured at the two dairy plants after 1, 15, 30, 60, 90, 120, 180, and 240 days of ripening are shown in Table 1. Significant differences were observed in DM ($p < 0.001$), pH ($p < 0.001$), fat ($p < 0.01$), and protein ($p < 0.001$) between the two dairy plants, while water activity was not affected. Regarding sampling zone, a_W and pH were significantly lower in the external part, whereas DM was significantly higher in the external as compared to the internal part. As expected, a_W decreased while DM increased significantly during cheese ripening. Percentages of fat and protein did not show differences due to ripening time or sampling zone.

Table 1. Mean values ± SD of chemical variables in raw sheep milk cheeses from two dairies (D1 and D2) sampled at external and internal zones during ripening.

Dairy Plant	Days of Ripening	pH		Water Activity (a_W)		Dry Matter (DM)		Fat (% DM)		Protein (% DM)		NaCl (% DM)	
		Internal	External	Internal	External	Internal	External	Internal	External	Internal	External	Internal	External
D1	1	5.46 ± 0.15	5.45 ± 0.16	0.98 ± 0.00	0.98 ± 0.01	54.3 ± 1.0	54.1 ± 1.6	54.5 ± 2.5	52.4 ± 1.9	37.7 ± 1.1	37.2 ± 0.6	1.27 ± 0.08	1.51 ± 0.29
	15	5.47 ± 0.09	5.46 ± 0.04	0.98 ± 0.00	0.97 ± 0.00	55.9 ± 5.1	54.6 ± 2.3	50.9 ± 3.3	51.6 ± 1.6	35.5 ± 3.9	35.5 ± 2.5	1.70 ± 0.13	3.55 ± 0.17
	30	5.48 ± 0.09	5.53 ± 0.08	0.97 ± 0.01	0.96 ± 0.01	55.1 ± 1.8	55.8 ± 0.4	52.8 ± 3.8	50.6 ± 3.5	37.0 ± 1.1	36.1 ± 1.5	1.92 ± 0.43	3.59 ± 0.33
	60	5.47 ± 0.12	5.52 ± 0.05	0.97 ± 0.00	0.95 ± 0.00	57.0 ± 0.7	58.9 ± 1.2	53.6 ± 2.2	51.7 ± 2.2	37.1 ± 0.1	36.0 ± 1.0	2.34 ± 0.32	3.63 ± 0.12
	90	5.45 ± 0.05	5.37 ± 0.05	0.96 ± 0.00	0.95 ± 0.00	59.3 ± 0.6	61.3 ± 1.7	52.3 ± 3.8	53.5 ± 5.9	37.2 ± 0.8	37.4 ± 1.6	2.36 ± 0.05	3.29 ± 0.07
	120	5.46 ± 0.11	5.41 ± 0.09	0.96 ± 0.00	0.95 ± 0.00	60.0 ± 1.1	62.8 ± 2.0	55.0 ± 5.6	53.0 ± 2.5	38.8 ± 3.3	36.3 ± 0.6	2.61 ± 0.32	3.09 ± 0.29
	180	5.38 ± 0.13	5.32 ± 0.16	0.94 ± 0.00	0.93 ± 0.01	63.6 ± 1.3	69.1 ± 2.9	55.6 ± 2.3	48.3 ± 3.5	36.3 ± 0.9	36.4 ± 0.8	3.15 ± 0.13	2.98 ± 0.26
	240	5.54 ± 0.11	5.43 ± 0.07	0.92 ± 0.00	0.91 ± 0.00	66.6 ± 1.1	71.8 ± 0.4	52.5 ± 2.6	52.8 ± 1.7	36.5 ± 0.6	36.6 ± 0.6	3.07 ± 0.26	2.92 ± 0.28
D2	1	5.63 ± 0.03	5.58 ± 0.03	0.98 ± 0.00	0.98 ± 0.01	57.5 ± 1.4	56.1 ± 1.5	51.8 ± 2.8	51.2 ± 2.4	40.1 ± 1.4	40.6 ± 1.0	0.64 ± 0.34	1.07 ± 0.51
	15	5.71 ± 0.11	5.77 ± 0.12	0.97 ± 0.01	0.98 ± 0.00	58.2 ± 2.0	58.1 ± 1.3	49.9 ± 2.3	50.0 ± 2.6	39.9 ± 2.6	39.1 ± 2.3	0.99 ± 0.44	2.03 ± 0.97
	30	5.73 ± 0.04	5.65 ± 0.07	0.97 ± 0.00	0.96 ± 0.01	59.2 ± 1.0	58.7 ± 1.1	50.7 ± 3.0	50.6 ± 2.2	39.7 ± 1.9	39.2 ± 1.8	1.1 ± 0.38	2.66 ± 0.54
	60	5.61 ± 0.08	5.58 ± 0.15	0.97 ± 0.00	0.95 ± 0.01	60.8 ± 1.3	63.2 ± 1.4	51.3 ± 2.6	50.3 ± 2.8	37.5 ± 3.8	39.5 ± 1.4	1.44 ± 0.50	2.57 ± 0.55
	90	5.75 ± 0.04	5.60 ± 0.01	0.96 ± 0.00	0.95 ± 0.01	62.8 ± 0.7	66.3 ± 2.0	51.0 ± 1.3	50.5 ± 2.5	39.9 ± 0.8	39.3 ± 1.3	1.38 ± 0.36	2.06 ± 0.27
	120	5.81 ± 0.18	5.61 ± 0.27	0.95 ± 0.01	0.94 ± 0.01	64.3 ± 0.9	69.0 ± 1.1	52.1 ± 3.9	49.1 ± 2.8	40.5 ± 0.6	38.5 ± 3.4	1.73 ± 0.59	2.09 ± 0.51
	180	5.82 ± 0.17	5.53 ± 0.13	0.94 ± 0.00	0.93 ± 0.01	67.2 ± 0.3	72.3 ± 0.6	52.9 ± 4.0	49.3 ± 2.0	38.7 ± 0.8	38.4 ± 0.5	1.8 ± 0.61	1.9 ± 0.59
	240	5.76 ± 0.03	5.59 ± 0.13	0.93 ± 0.00	0.91 ± 0.01	68.1 ± 0.2	74.4 ± 1.0	48.8 ± 0.2	48.4 ± 0.0	40.6 ± 3.8	38.6 ± 1.7	2.21 ± 0.22	2.21 ± 0.57
Dairy plant (D)		***		NS		***		**		***		***	
Sampling zone (SZ)		**		***		***		NS		NS		***	
Ripening time		NS		***		***		NS		NS		*	

Significance levels: ***, $p < 0.001$; **, $p < 0.01$; * $p < 0.05$; NS, not significant; SD, standard deviation; NaCl, sodium chloride.

The artisanal cheesemaking process is one of the main reasons for the variations observed between dairies. Some differences in milk management before renneting, curd cutting, curd temperature, and cheesemaking technology are believed to be potentially responsible for different physicochemical composition in cheeses made under similar conditions [21]. In addition, the brining conditions could lead to significant differences in dry matter content, corresponding to different moisture levels of the cheeses, as reported by Todaro et al. [22] for Italian raw sheep milk cheeses.

The pH values ranged from 5.32 to 5.82 and significant differences between dairy plants were found, with cheeses from D1 showing lower pH values than those from D2. Likewise, the external parts of the cheeses showed lower pH values than the internal parts. Few studies have reported pH values in different parts of cheese. Malacarne et al. [7] and Sihufe et al. [9] did not find differences between zones in Parmigiano-Reggiano cheeses for this parameter. However, in agreement with our data, Tosi et al. [8] found higher pH values in the internal parts of Parmigiano-Reggiano than in external parts. Regardless of the different results found in other studies, our results fall within the range of pH allowed by PDO regulations of the European Commission for Spanish sheep milk cheeses such as Idiazábal (pH 5.1–5.8), La Serena (pH 5.2–5.9), Manchego (pH 4.8–5.8), Zamorano (pH 5.1–5.8), and Torta del Casar (pH 5.2–5.9).

At the beginning of the ripening period, there were no differences in water activity between the external and internal parts of the cheeses (0.98 initial value). However, throughout ripening, the external zone presented significantly lower a_W values than the internal zone. Pellegrino et al. [23] found the same phenomenon in Grana Padano cheese. Water activity value decreased 0.07 units (from 0.98 to 0.91) during 240 days of ripening, in agreement with other Spanish sheep milk cheeses such as Manchego [24], Idiazábal [25], and Los Pedroches [26].

The most important differences due to sampling zone were observed in dry matter and salt content, as reported in other cheese varieties such as Reggianito Argentino [9]. When cheeses are salted by brine immersion, salt is taken up while moisture is simultaneously lost. Comparing external and internal zones, no differences were found in dry matter levels at the beginning of ripening (55.5% on average). However, the differences in DM were gradually noticeable during ripening due to the natural and progressive loss of moisture, until reaching a 5–6 percentage unit difference at the end of the 240 days. These results contrast with other studies in which differences in dry matter between the external and internal zones were evident from the first stage of ripening. Thus, in Parmigiano-Reggiano cheese, the difference between internal and external zones was around 2–4% [7], and in Reggianito Argentino cheese it was 8–10% [9,27]. These varieties of cheese remain in brine for about 25 days and 9 days, respectively. However, our samples were kept in brine for only 12–24 h. Other varieties, such as Port Salut Argentino cheese packaged with plastic film, also showed a constant difference (4%) in DM between sampling zones, which remained constant throughout ripening [10].

In Grana Padano cheese [23] the DM difference between the center and the periphery of the cheese was 6%, similar to the difference observed between the parts analyzed in our study at 240 days of ripening. Although fat and protein contents were different in cheeses depending on the dairy plant, no significant differences in these parameters were observed between sampling zones during ripening.

3.2. Salt Distribution

As expected, NaCl distribution changed during ripening. Figure 2 shows the NaCl content (g NaCl/100 g DM) in the cheese samples from both dairies in the internal and external zones during 240 days of ripening. At the beginning of maturation, the NaCl content in the outer zone (1.51% and 1.07% in D1 and D2, respectively) was significantly higher than in the inner zone (1.27% and 0.64% in D1 and D2, respectively) ($p < 0.001$). Regardless of the differences in final salt content at 240 ripening days (3.00% and 2.20% in D1 and D2, respectively), the phenomenon of salt diffusion from the outer zone into the cheese evolved similarly during ripening. Thus, homogeneous salt distribution was reached at 180 days of ripening in cheeses subjected to the two brining conditions. This suggests that the salt diffusion phenomenon occurred very consistently and was not adversely affected by the multilobed shape of the cheese.

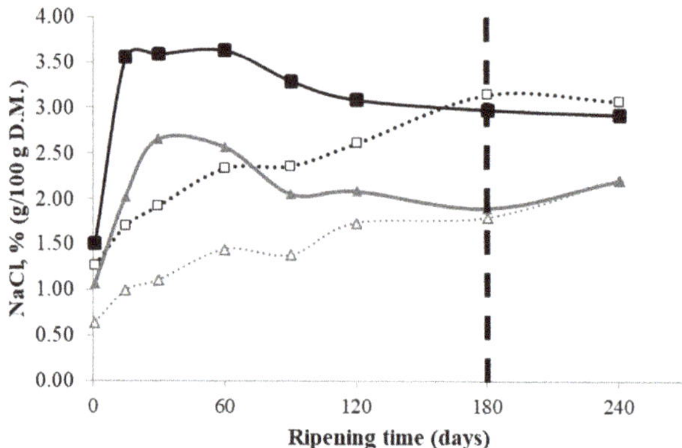

Figure 2. Sodium chloride content (g NaCl/100 g DM) in raw sheep milk cheeses made at two dairies (D1 and D2) during 240 days of ripening. (···□···): D1 internal zone, (-■-): D1 external zone, (···△···): D2 internal zone, (—): D2 external zone.

Establishing the salt equilibrium in ripened cheese is a very slow process that can be affected by many factors, such as dairy plant, milk batch, brining conditions, and other cheesemaking processes [28]. The pH could have played a role in the final salt content of the cheeses, as higher pH values contribute to decreasing the negative charge of the casein micelles, leading to less retention of salt. In our case, the cheeses from the two dairies differed in their internal pH (5.38 in D1 versus 5.82 in D2 at 180 days of ripening), which corresponded to final salt contents of 3.0% and 2.2%, respectively. Additionally, cheeses with higher fat content generally show reduced rates of salt absorption, because the fat globules may obstruct the ducts where the brine enters the cheese. This phenomenon did not happen in our cheeses, since those with more fat content (55.6% in D1 versus 52.9% in D2) also reached a higher final salt concentration.

Although cheeses are manufactured under similar procedures, it is widely accepted that there are variations in the sensory attributes and compositional characteristics, mainly due to cheesemaking practices applied at each dairy [29]. In this study, cheeses salted in brine at 16°Baumé for 24 h had a 1% higher salt content than cheeses brined at 22°Baumé for 12 h.

3.3. Proteolysis

3.3.1. Nitrogen Fractions

The nitrogen fractions studied changed as a function of ripening time. Figure 3A,B reflect the extent of proteolysis (WSN/TN and pH 4.4-SN/TN), Figure 3C represents TCA-SN/TN as ripening depth index, and Figure 3D shows PTA-SN/TN as free amino acid index. The formation of soluble nitrogen compounds during cheese ripening is an index of the rate and extent of proteolysis, in that it is an indicator of casein hydrolysis brought about by the action of rennet and milk proteases present at the beginning of ripening. The extent of proteolysis during the studied period was moderate, as expected in raw sheep milk cheese made with calf rennet. WSN/TN reached a final value of around 25% total nitrogen.

These results are in agreement with those reported for similar varieties of raw sheep milk cheeses such as Manchego [24], Idiazábal [25], and Roncal [30]. The level of soluble nitrogen for each fraction increased significantly during the 240 days of ripening, while free amino acid content (PTA-SN/TN) decreased after 180 days of ripening. Several FAAs undergo transamination, decarboxylation, and deamination reactions rather extensively in late ripening [31]. This suggests that free amino acids enter other metabolic pathways from 180 days of ripening.

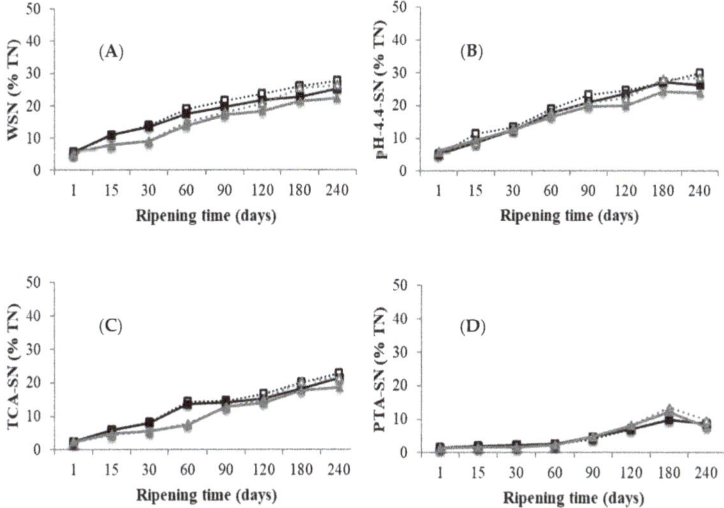

Figure 3. Soluble nitrogen fractions in raw sheep milk cheeses made at two dairies (D1 and D2) during 240 days of ripening (% of total nitrogen). (⋯□⋯): D1 internal zone, (–■–): D1 external zone, (⋯△⋯): D2 internal zone, (–×–): D2 external zone. (**A**) water-soluble nitrogen (WSN)/total nitrogen (TN); (**B**) pH 4.4 soluble nitrogen (SN)/TN; (**C**) trichloroacetic acid 12% (*w/v*) (TCA)-SN/TN; (**D**) phosphotungstic acid (PTA)-SN/TN.

Significant differences ($p < 0.05$) between dairy plants were observed for the nitrogen fractions studied (WSN, pH 4.4-SN, TCA-SN, and PTA-SN). Cheeses made at D1 showed higher SN values for each nitrogen fraction than D2, while cheeses from D2 had lower salt content than those from D1. These results are in contrast to the inverse relationship between the extent of proteolysis and salt concentration widely reported in cheese [2,31]. Our results differed from those reported by other authors who studied other cheesemaking procedures with different amounts of rennet or dosages of starter cultures. Bustamante et al. [32] found that the amount of rennet had a significant effect ($p < 0.01$) on the extent of proteolysis (WSN/NT), and Galán et al. [33] concluded that the amount of rennet had more effect on proteolysis than the type of rennet used. Considering these results, the cheesemaking process has a greater influence on the degree of proteolysis than the salt concentration.

Regarding the sampling zone factor, no significant differences ($p > 0.05$) were found on the ripening index except for samples from dairy plant D2, which presented somewhat lower WSN/TN and pH 4.4-SN/TN at the external zone after 120 days of ripening. This agrees with the findings of Sihufe et al. [27], who reported that the degradation of casein was larger in the central zone than the external zone in Reggianito Argentino cheese, which may be related to the fact that the enzymatic system is favored by higher moisture content and lower salt content. Despite the differences found in salt concentration between the external and internal zones of the cheese until 180 days, the proteolysis index did not differ significantly between sampling zones.

3.3.2. Free Amino Acids

The major individual amino acids were glutamic acid (Glu), lysine (Lys), asparagine (Asn), serine (Ser), ornithine (Orn), leucine (Leu), and phenylalanine (Phe), representing around 60% of the total free amino acids in the later stages of ripening (>90 days) (data not shown). Cysteine was not detected in any cheese samples and tryptophan was detected only in samples from D1 dairy.

Arginine concentration decreased later in ripening. This reduction is related to the metabolism of nonstarter lactic acid bacteria (NSLAB), which are able to convert arginine to citrulline and ornithine (not a constituent of casein) via the deiminase pathway [34]. This could be considered a positive aspect, as arginine is associated with bitter taste [35]. The effect of the sampling zone was not significant ($p > 0.05$) for most FAAs. Nevertheless, in the current samples, ornithine and arginine were inversely affected by sampling zone: while ornithine increased in the internal zone of cheeses during ripening, arginine concentration decreased. This could be associated with a higher growth of mesophilic lactobacilli in the internal zone due to lower salt content and higher water activity than in the external zone.

Gamma-aminobutyric acid (GABA), a nonprotein amino acid generated through decarboxylation of glutamic acid, has been correlated with an increased number of cheese eyes, although it has no direct or indirect impact on cheese flavor [36]. In this study, GABA values were around 3% of the total FAA at 240 days, which is similar to other raw sheep milk cheeses such as Idiazábal [37] and Manchego [38]. A higher concentration of GABA found in the internal zone could be related to the action of mesophilic lactobacilli [35]. Tyrosine was affected by sampling zone and its concentration was lower in the internal zone. In addition, its relative concentration decreased throughout ripening.

Total FAAs increased significantly during ripening, coinciding with concomitant increases in the concentrations of most individual FAAs, except for arginine and taurine. Although the sampling zone effect was not significant ($p > 0.05$) for most individual FAAs, significant differences were found for Arg, Tyr, and Orn at both dairies, and for GABA at D1. The Pearson correlation coefficient was calculated to estimate the relationship between individual amino acid concentrations and ripening time (Supplementary Table S2).

Principal component analysis (PCA) of proteolysis data (nitrogen fractions, individual and total FAA concentrations) distributed the samples according to ripening time and dairy plant. Two principal components, PC1 (73.51%) and PC2 (7.34%), explained 80.85% of the variance (Figure 4). The score plot organizes samples with long ripening (>60 days) according to dairy plant. However, younger cheeses (<60 days) were grouped together. PC1 presented high correlation with most individual amino acids (apart from Tau, Arg, and Tyr), while PC2 showed the highest correlations with SN-pH4.4/NT, WSN/NT, and leucine content.

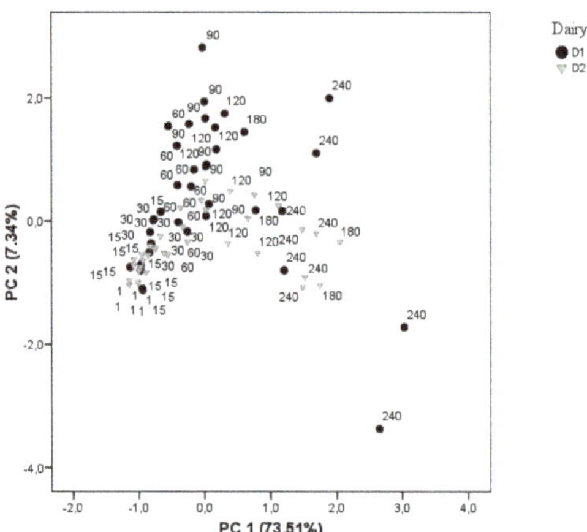

Figure 4. Plot of first two principal components obtained from principal component analysis of proteolysis data of cheeses from two dairies (D1 and D2) during ripening (1, 15, 30, 60, 90, 120, 180, and 240 days).

3.4. Lipolysis

The extent of lipolysis was moderate as expected in semi-hard/hard cheese varieties due to low lipolytic activity of lactic acid bacteria [39]. Total FFA values expressed as the sum of individual FFAs during ripening are shown in Figure 5. Total FFA levels increased significantly ($p < 0.05$) during ripening in samples from both dairies and were significantly higher ($p < 0.01$) in samples from D1 (~3000 mg FFA/kg of cheese) than from D2 (~2250 mg FFA/kg of cheese) at the end of ripening. However, no significant variation in the composition of total FFA was observed in the two sampling zones. Differences in FFA concentration may be significant even in the same cheese variety. The intra-variety difference in total FFA has been related to milk composition and microbial activity, depending on sheep breed, stage of lactation, season, and feeding, and many studies on cheeses manufactured at different dairy plants corroborate this variability [40].

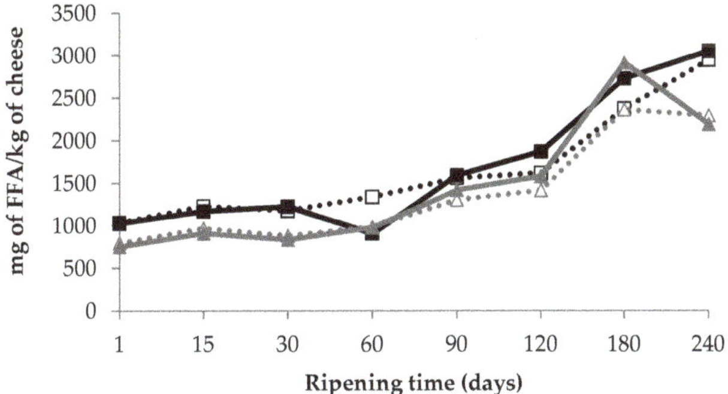

Figure 5. Total free fatty acid (FFA) values expressed as sum of individual FFAs in raw sheep milk cheeses made at two dairies (D1 and D2) during 240 days of ripening. (···□··): D1 internal zone, (—■—): D1 external zone, (···△··): D2 internal zone, (—▲—): D2 external zone.

The comparison between sampling zones showed a uniform distribution of FFAs for Reggianito Argentino at 4 months of ripening, in agreement with the data reported by Sihufe et al. [9]. However, Malacarne et al. [7] found higher FFA content in the external zone of Parmigiano-Reggiano cheese at 24 months of ripening. Samples with higher salt content (D1) showed a greater extent of lipolysis. Previous studies in Cheddar cheese manufactured with pasteurized milk indicated that NaCl may inhibit lipolysis [41]. However, other investigations in Cheddar cheese concluded that FFAs were unaffected by salt reduction [6]. This suggests that the FFA values in our study could be ascribed to different milk components such as lipoprotein lipase (LPL), the principal indigenous lipase in raw milk, as well as to microbial growth [39].

The individual FFAs were grouped into short-chain fatty acids (C4:0–C8:0, SCFAs), medium-chain fatty acids (C10:0–C14:0, MCFAs), and long-chain fatty acids (C16:0–C18:3, LCFAs), and were expressed as percentages of total FFA. The initial proportion of LCFA (65% of total FFA) decreased during ripening to 45%. LCFAs have a higher perception threshold and do not contribute much to cheese flavor. Relative concentration of MCFA (25–30%) did not show significant variation during ripening. However, SCFA proportion increased during the study period from 10 to 25%. It is usually considered that short-chain fatty acids have a significant impact on the development of the characteristic cheese flavor [39]. Our results show a preference of lipases for short-chain fatty acid residues, since these acids predominate at the sn-3 position. This has been reported in other cheeses such as Idiazábal [19], Manchego [24], and Queso Picante [42].

As a result of lipolysis, all studied individual FFA concentrations increased significantly ($p < 0.001$) with ripening time. The major FFAs were palmitic (C16:0) and oleic (C18:1) acids, coinciding with other sheep milk cheeses [43]. Pearson correlation coefficients between the percentage of FFA and ripening time were calculated (Supplementary Table S3). The proportions of butyric (C4:0) and caproic (C6:0) acids, which are known to contribute actively to cheese flavor and aroma, showed the highest correlation coefficient ($r > 0.80$). Despite the statistical differences found for some FFAs among dairy plants, all samples showed a similar pattern of lipolysis.

4. Conclusions

In this artisanal hard cheese, salt was homogeneously distributed in the entire product at 180 days of ripening. The multilobed shape of the cheese did not adversely affect the rate of salt absorption. The brining conditions (time and salinity) had no influence on the rate of salt penetration. However, they had a significant effect on NaCl content, as cheeses salted in brine solution at 16°Baumé for 24 h showed a higher salt content than cheeses brined at 22°Baumé for 12 h. Proteolysis and lipolysis showed similar patterns in all samples and evolved homogeneously during the ripening period, in both the internal and external zones of cheese. These findings may constitute useful information for cheese producers in deciding when to market this product in order to reach uniformity in salt distribution.

Supplementary Materials: The following are available online at http://www.mdpi.com/2304-8158/8/3/100/s1: Table S1. Quaternary gradient for separation of free amino acids in cheese derivatized with 6-aminoquinolyl-N-hydroxysuccinimidyl carbamate (AQC) by high-performance liquid chromatography (HPLC-DAD-FLD). Table S2. Significance of ripening time, dairy plant, and sampling zone on individual concentration of free amino acids (FFAs) and total FAA, and Pearson coefficient (r) in raw sheep milk cheeses made in two dairies (D1 and D2) during 240 days of ripening. Table S3. Significance of ripening time, dairy plant, and sampling zone on relative concentrations of individual free fatty acids (FFAs) and Pearson coefficient (r) in raw sheep milk cheeses made in two dairies (D1 and D2) during 240 days of ripening.

Author Contributions: Conceptualization, O.E., T.J. and A.A.; Methodology, validation, formal analysis, data curation, writing the original draft of the article and visualization, O.E.; Funding acquisition, project administration, resources, T.J.; Supervision and conducted review of the article, T.J. and A.A.

Funding: This research was funded by the Instituto Nacional de Investigación y Tecnología Agraria y Alimentaria (INIA) (BOE-A-2016-1448) and Gobierno de Aragón-FEDER (Grupo A06_17R).

Conflicts of Interest: The authors declare no conflict of interest.

References

1. Montel, M.C.; Buchin, S.; Mallet, A.; Delbes-Paus, C.; Vuitton, D.A.; Desmasures, N.; Berthier, F. Traditional cheeses: Rich and diverse microbiota with associated benefits. *Int. J. Food Microbiol.* **2014**, *177*, 136–154. [CrossRef] [PubMed]
2. Guinee, T.P.; Fox, P.F. *Salt in Cheese: Chemistry, Physics and Microbiology. General Aspects*, 3rd ed.; Chapman & Hall: London, UK, 2004; Volume 1, pp. 207–259.
3. McCarthy, C.M.; Wilkinson, M.G.; Kelly, P.M.; Guinee, T.P. Effect of salt and fat reduction on the composition, lactose metabolism, water activity and microbiology of Cheddar cheese. *Dairy Sci. Technol.* **2015**, *95*, 587–611. [CrossRef]
4. Baptista, D.P.; da Silva Araújo, F.D.; Eberlin, M.N.; Gigante, M.L. Reduction of 25% salt in Prato cheese does not affect proteolysis and sensory acceptance. *Int. Dairy J.* **2017**, *75*, 101–110. [CrossRef]
5. Zabaleta, L.; Albisu, M.; Ojeda, M.; Gil, P.; Etaio, I.; Perez-Elortondo, F.; de Renobales, M.; Barron, L. Occurrence of sensory defects in semi-hard ewe's raw milk cheeses. *Dairy Sci. Technol.* **2016**, *96*, 53–65. [CrossRef]
6. McCarthy, C.M.; Kelly, P.M.; Wilkinson, M.G.; Guinee, T.P. Effect of fat and salt reduction on the changes in the concentrations of free amino acids and free fatty acids in Cheddar-style cheeses during maturation. *J. Food Compos. Anal.* **2017**, *59*, 132–140. [CrossRef]
7. Malacarne, M.; Summer, A.; Franceschi, P.; Formaggioni, P.; Pecorari, M.; Panari, G.; Mariani, P. Free fatty acid profile of Parmigiano-Reggiano cheese throughout ripening: Comparison between the inner and outer regions of the wheel. *Int. Dairy J.* **2009**, *19*, 637–641. [CrossRef]
8. Tosi, F.; Sandri, S.; Tedeschi, G.; Malacarne, M.; Fossa, E. Variazioni di composizione e proprietà fisico-chimiche del Parmigiano-Reggiano durante la maturazione e in differenti zone della forma. *Scienza e Tecnica Lattiero-Casearia.* **2008**, *59*, 507–528.

9. Sihufe, G.A.; Zorrilla, S.E.; Mercanti, D.J.; Perotti, M.C.; Zalazar, C.A.; Rubiolo, A.C. The influence of ripening temperature and sampling site on the lipolysis in Reggianito Argentino cheese. *Food Res. Int.* **2007**, *40*, 1220–1226. [CrossRef]
10. Verdini, R.A.; Zorrilla, S.E.; Rubiolo, A.C. Characterisation of soft cheese proteolysis by RP-HPLC analysis of its nitrogenous fractions. Effect of ripening time and sampling zone. *Int. Dairy J.* **2004**, *14*, 445–454. [CrossRef]
11. ISO. *Determination of the Total Solid Content (Reference Method)*; International Organization for Standardization: Geneva, Switzerland, 2004.
12. ISO. *Cheese and Processed Cheese Products. Determination of Fat Content. Gravimetric Method*; International Organization for Standardization: Geneva, Switzerland, 2004.
13. ISO. *Determination of Nitrogen Content. Part 1: Kjeldahl Principle and Crude Protein Calculation*; International Organization for Standardization: Geneva, Switzerland, 2014.
14. ISO. *Determination of Chloride Content. Potentiometric Titration Method*; International Organization for Standardization: Geneva, Switzerland, 2006.
15. Butikofer, U.; Ruegg, M.; Ardo, Y. Determination of Nitrogen Fractions in Cheese-Evaluation of a Collaborative Study. *LWT Food Sci. Technol.* **1993**, *26*, 271–275. [CrossRef]
16. Milesi, M.M.; Vinderola, G.; Sabbag, N.; Meinardi, C.A.; Hynes, E. Influence on cheese proteolysis and sensory characteristics of non-starter lactobacilli strains with probiotic potential. *Food Res. Int.* **2009**, *42*, 1186–1196. [CrossRef]
17. Waters, AccQ. *Tag Chemistry Package: Instruction Manual*; Waters Corporation: Milford, MA, USA, 2008.
18. Estrada, O.; Molino, F.; Sanz, M.A.; Juan, T. Optimization of a Chromatographic Method for the Determination of Free Amino Acids in Cheese. In Proceedings of the VI IBEROLAB Virtual Ibero-American Conference on Quality Management in Laboratories. 2011. Available online: http://www.iberolab.org (accessed on 31 January 2019).
19. Chavarri, F.; Virto, M.; Martin, C.; Najera, A.I.; Santisteban, A.; Barron, L.J.R.; De Renobales, M. Determination of free fatty acids in cheese: Comparison of two analytical methods. *J. Dairy Res.* **1997**, *64*, 445–452. [CrossRef]
20. De Jong, C.; Badings, H. Determination of Free Fatty Acids in Milk and Cheese—Procedures for Extraction, Clean Up, and Capillary Gas-Chromatographic Analysis. *HRC-J. High Resolut. Chromatogr.* **1990**, *13*, 94–98. [CrossRef]
21. Mucchetti, G.; Bonvini, B.; Remagni, M.C.; Ghiglietti, R.; Locci, F.; Barzaghi, S.; Francolino, S.; Perrone, A.; Rubiloni, A.; Campo, P.; et al. Influence of cheese-making technology on composition and microbiological characteristics of Vastedda cheese. *Food Control* **2008**, *19*, 119–125. [CrossRef]
22. Todaro, M.; Francesca, N.; Reale, S.; Moschetti, G.; Vitale, F.; Settanni, L. Effect of different salting technologies on the chemical and microbiological characteristics of PDO Pecorino Siciliano cheese. *Eur. Food Res. Technol.* **2011**, *233*, 931–940. [CrossRef]
23. Pellegrino, L.; Battelli, G.; Resmini, P.; Ferranti, P.; Barone, F.; Addeo, F. Effects of heat load gradient occurring in moulding on characterization and ripening of Grana Padano. *Dairy Sci. Technol.* **1997**, *77*, 217–228. [CrossRef]
24. Poveda, J.M.; Sousa, M.J.; Cabezas, L.; McSweeney, P.L.H. Preliminary observations on proteolysis in Manchego cheese made with a defined-strain starter culture and adjunct starter (*Lactobacillus plantarum*) or a commercial starter. *Int. Dairy J.* **2003**, *13*, 169–178. [CrossRef]
25. Mendia, C.; Ibañez, F.C.; Torre, P.; Barcina, Y. Influence of the season on proteolysis and sensory characteristics of Idiazabal cheese. *J. Dairy Sci.* **2000**, *83*, 1899–1904. [CrossRef]
26. Sanjuan, E.; Millan, R.; Saavedra, P.; Carmona, M.A.; Gome, R.; Fernandez-Salguero, J. Influence of animal and vegetable rennet on the physicochemical characteristics of Los Pedroches cheese during ripening. *Food Chem.* **2002**, *78*, 281–289. [CrossRef]
27. Sihufe, G.A.; Zorrilla, S.E.; Rubiolo, A.C. The influence of ripening temperature and sampling site on the proteolysis in Reggianito Argentino cheese. *LWT-Food Sci. Technol.* **2010**, *43*, 247–253. [CrossRef]
28. Akkerman, M.; Kristensen, L.S.; Jespersen, L.; Ryssel, M.B.; Mackie, A.; Larsen, N.N.; Andersen, U.; Nørgaard, M.K.; Løkke, M.M.; Møller, J.R. Interaction between sodium chloride and texture in semi-hard Danish cheese as affected by brining time, DL-starter culture, chymosin type and cheese ripening. *Int. Dairy J.* **2017**, *70*, 34–45. [CrossRef]

29. Pintado, A.I.E.; Pinho, O.; Ferreira, I.M.P.L.V.O.; Pintado, M.M.E.; Gomes, A.M.P.; Malcata, F.X. Microbiological, biochemical and biogenic amine profiles of Terrincho cheese manufactured in several dairy farms. *Int. Dairy J.* **2008**, *18*, 640. [CrossRef]
30. Irigoyen, A.; Izco, J.M.; Ibañez, F.C.; Torre, P. Influence of calf or lamb rennet on the physicochemical, proteolytic, and sensory characteristics of an ewe's-milk cheese. *Int. Dairy J.* **2002**, *12*, 27–34. [CrossRef]
31. Fox, P.F.; Guinee, T.P.; Cogan, T.M.; McSweeney, P.L.H. *Fundamentals of Cheese Science*; Aspen Publisher Inc.: Gaithersburg, MD, USA, 2000.
32. Bustamante, M.A.; Virto, M.; Aramburu, M.; Barron, L.J.R.; Perez-Elortondo, F.J.; Albisu, M.; de Renobales, M. Lamb rennet paste in ovine cheese (Idiazabal) manufacture. Proteolysis and relationship between analytical and sensory parameters. *Int. Dairy J.* **2003**, *13*, 547–557. [CrossRef]
33. Galán, E.; Prados, F.; Pino, A.; Tejada, L.; Fernández-Salguero, J. Influence of different amounts of vegetable coagulant from cardoon *Cynara cardunculus* and calf rennet on the proteolysis and sensory characteristics of cheeses made with sheep milk. *Int. Dairy J.* **2008**, *18*, 93–98. [CrossRef]
34. Laht, T.; Kask, S.; Elias, P.; Adamberg, K.; Paalme, T. Role of arginine in the development of secondary microflora in Swiss-type cheese. *Int. Dairy J.* **2002**, *12*, 831–840. [CrossRef]
35. Poveda, J.M.; Chicón, R.; Cabezas, L. Biogenic amine content and proteolysis in Manchego cheese manufactured with *Lactobacillus paracasei* subsp. *paracasei* as adjunct and other autochthonous strains as starters. *Int. Dairy J.* **2015**, *47*, 94–101. [CrossRef]
36. Christensen, J.E.; Dudley, E.G.; Pederson, J.A.; Steele, J.L. Peptidases and amino acid catabolism in lactic acid bacteria. *Antonie Van Leeuwenhoek* **1999**, *76*, 217–246. [CrossRef] [PubMed]
37. Oneca, M.; Ortigosa, M.; Irigoyen, A.; Torre, P. Proteolytic activity of some *Lactobacillus paracasei* strains in a model ovine-milk curd system: Determination of free amino acids by RP-HPLC. *Food Chem.* **2007**, *100*, 1602–1610. [CrossRef]
38. Muñoz, N.; Ortigosa, M.; Torre, P.; Izco, J.M. Free amino acids and volatile compounds in an ewe's milk cheese as affected by seasonal and cheese-making plant variations. *Food Chem.* **2003**, *83*, 329–338. [CrossRef]
39. Collins, Y.F.; McSweeney, P.L.H.; Wilkinson, M.G. Lipolysis and free fatty acid catabolism in cheese: A review of current knowledge. *Int. Dairy J.* **2003**, *13*, 841–866. [CrossRef]
40. Boutoial, K.; Alcántara, Y.; Rovira, S.; García, V.; Ferrandini, E.; López, M.B. Influence of ripening on proteolysis and lipolysis of Murcia al Vino cheese. *Int. J. Dairy Technol.* **2013**, *66*, 366–372. [CrossRef]
41. Reddy, K.A.; Marth, E.H. Composition of Cheddar cheese made with sodium chloride and potassium chloride either singly or as mixtures. *J. Food Compos. Anal.* **1993**, *6*, 354–363. [CrossRef]
42. Freitas, A.C.; Fresno, J.M.; Prieto, B.; Malcata, F.X.; Carballo, J. Effects of ripening time and combination of ovine and caprine milks on proteolysis of Picante cheese. *Food Chem.* **1997**, *60*, 219–229. [CrossRef]
43. Poveda, J.M.; Perez-Coello, M.S.; Cabezas, L. Seasonal variations in the free fatty acid composition of Manchego cheese and changes during ripening. *Eur. Food Res. Technol.* **2000**, *210*, 314–317. [CrossRef]

Article

Ripening of Hard Cheese Produced from Milk Concentrated by Reverse Osmosis

Anastassia Taivosalo [1,2,*], Tiina Kriščiunaite [1], Irina Stulova [1], Natalja Part [1], Julia Rosend [1,2], Aavo Sõrmus [1] and Raivo Vilu [1,2]

1 Center of Food and Fermentation Technologies, Akadeemia tee 15A, 12618 Tallinn, Estonia; tiina@tftak.eu (T.K.); irina.stulova@tftak.eu (I.S.); natalja@tftak.eu (N.P.); julia@tftak.eu (J.R.); aavo@tftak.eu (A.S.); raivo@tftak.eu (R.V.)

2 Department of Chemistry and Biotechnology, Tallinn University of Technology, Akadeemia tee 15, 12618 Tallinn, Estonia

* Correspondence: anastassia@tftak.eu

Received: 17 April 2019; Accepted: 13 May 2019; Published: 15 May 2019

Abstract: The application of reverse osmosis (RO) for preconcentration of milk (RO-milk) on farms can decrease the overall transportation costs of milk, increase the capacity of cheese production, and may be highly attractive from the cheese manufacturer's viewpoint. In this study, an attempt was made to produce a hard cheese from RO-milk with a concentration factor of 1.9 (RO-cheese). Proteolysis, volatile profiles, and sensory properties were evaluated throughout six months of RO-cheese ripening. Moderate primary proteolysis took place during RO-cheese ripening: about 70% of α_{s1}-casein and 45% of β-casein were hydrolyzed by the end of cheese maturation. The total content of free amino acids (FAA) increased from 4.3 to 149.9 mmol kg^{-1}, with Lys, Pro, Glu, Leu, and γ-aminobutyric acid dominating in ripened cheese. In total, 42 volatile compounds were identified at different stages of maturation of RO-cheese; these compounds have previously been found in traditional Gouda-type and hard-type cheeses of prolonged maturation. Fresh RO-cheese was characterized by a milky and buttery flavor, whereas sweetness, saltiness, and umami flavor increased during ripening. Current results prove the feasibility of RO-milk for the production of hard cheese with acceptable sensory characteristics and may encourage further research and implementation of RO technology in cheese manufacture.

Keywords: reverse osmosis; concentrated milk; hard cheese; cheese ripening; volatile compounds

---

1. Introduction

The application of a membrane separation technology has gained increasing attention in the dairy industry. The substantial continuing increase in the global milk and cheese production [1] calls for innovations in cheese milk pretreatment to improve the efficiency of cheese manufacturing plants retaining high quality of cheese. Preconcentration of large volumes of raw milk prior to transportation to a cheese plant reduces the delivery costs of milk, and potentially increases the plant capacity and cheese yields.

Ultra- (UF) and microfiltration (MF) for the concentration of cheese milk are the most widely and successfully applied techniques in cheesemaking [2] and have been of enormous research interest. UF retentates have been used to standardize milk protein content by entrapping the whey proteins into the cheese curd matrix to improve the cheese yield and composition of Mozzarella, Cheddar, Camembert, and Brie cheeses [3,4], and semihard cheese made from a mixture of ewes' and cows' milk [5]. Fewer studies have been performed on UF-Feta cheese, where the cheese rheological [6] and microstructural changes in fat globules during ripening [7] as well as accelerating the cheese ripening by the addition of lipase have been evaluated [8]. MF retentates with increased casein (CN) content can be added to

UF retentates to optimize the cheesemaking process by improving rennet coagulability, which results in a firmer gel and increased cheese yield [2,9]. CN enrichment by MF has been applied to produce Mozzarella [10], Cheddar [11,12], and semihard [13] cheeses with a good sensory quality. The effect of both MF and UF on Edam cheese yield and ripening has been evaluated by Heino et al. [14].

Reverse osmosis (RO) is a filtration method that separates the solutes with a molecular weight of 150 Da and less [2]. Thus, water passes through the membrane, while fats, proteins, lactose, and minerals are retained [2]. The process of RO in the dairy industry has been developed during the last 25 years [15,16] and has received considerable attention, especially that of whey protein concentration [17,18]. Moreover, this dewatering technology has been used for the concentration of skim milk before drying in the production of milk powders [19], and prior to yoghurt manufacture [20]. However, only a limited number of studies have demonstrated the application of RO in cheese technology. Agbevavi et al. [21] made the first attempt to produce Cheddar cheese from whole milk concentrated by RO (RO-milk) on a pilot plant scale. The texture of cheese was assumed to be not uniform because of the high lactose content of the retentate. Moreover, the unacceptable bacterial contamination of the final cheese was observed due to the poor sanitary conditions of the RO system [21]. Barbano and Bynum [22] evaluated the effect of different water reduction levels of whole milk obtained by RO on Cheddar cheese produced in a commercial cheese plant. The authors succeeded in the manufacture of the cheese with increased yield by using conventional cheesemaking equipment. The CN breakdown in Cheddar cheese produced from RO-milk has been shown to be similar to that of the control cheese at earlier stages of ripening, while the slower proteolysis has been noted in aged cheese [23]. Additionally, a significantly higher lactose content in Cheddar cheese produced from RO-milk compared to the cheese from unconcentrated milk was determined, which could cause increased lactic acid fermentation, affecting the sensory properties of cheese. Hydamaka et al. [24] used the whole milk RO retentate to produce the direct acidified cheese with a higher cheese yield and good sensory characteristics. At the moment, there have been no published data on the production and ripening of hard-type cheeses from RO-milk.

In this research, a concentrated milk was manufactured using RO filtration technology. The aim of the study was to produce a hard cheese from the RO-milk (RO-cheese) and to study the impact of milk concentration on cheese ripening. Both mesophilic and thermophilic starter cultures and high curd scalding temperature (48 °C) were used in RO-cheese production. The breakdown of caseins, release of amino acids and volatile compounds, as well as sensory properties of RO-cheese were evaluated during six months of ripening.

2. Materials and Methods

2.1. RO-Cheese Manufacture and Sampling

A Gouda-type cheese technology with certain modifications into the manufacturing protocol (i.e., addition of thermophilic starter cultures and high scalding temperature) was applied to obtain a cheese with hard texture and low moisture content. A hard cheese was produced in the pilot plant at the School of Service and Rural Economics (Olustvere, Estonia) from 100 L of whole bovine RO-milk (concentration factor of 1.9, dry matter 21.7% w/w, pH 6.44) pasteurized at 74 °C for 15 s. After heat treatment, the RO-milk was cooled to 32 °C and inoculated with mixed starter culture (10 U 100 L^{-1}) 30 min prior to addition of microbial rennet (8 g 100 L^{-1}; 1300 IMCU g^{-1}, Chymax, Chr. Hansen Ltd., Hørsholm, Denmark). The starter consisted of multiple strains of mesophilic and thermophilic lactic acid bacteria (LAB)—FD-DVS CHN-11, FD-DVS LH-B02, and FD-DVS ST-B01 (Chr. Hansen Ltd., Hørsholm, Denmark) mixed in the proportion 10:5:1, respectively—*Lactococcus lactis* subsp. *lactis*, *Lc. lactis* subsp. *cremoris*, *Lc. lactis* subsp. *diacetylactis*, *Leuconostoc* sp., *Streptococcus thermophilus*, and *Lactobacillus helveticus*. After coagulation, the curd was cut into 0.7 cm × 0.7 cm × 0.7 cm cubes and then slowly stirred for 20 min followed by removing 20 L of whey. Warm water was added to the vat and cheese grains were continuously mixed and heated at 48 °C for 30 min. After pre-pressing at 0.5 bar for

10 min and 1 bar for 10 min in the vat under the whey, the whey was drained off and pre-pressed grain cubes were transferred into low cylinder cheese molds. The cheeses were pressed three times for 20 min at 1, 1.8, and 2.1 bar, and brine-salted (20% NaCl, *w/v*) for 16 h at room temperature. Approximately 1 kg wheel-shaped cheeses were coated with wax (6.6 g kg^{-1} of cheese; Ceska®-coat WL01; CSK Food Enrichment, Netherlands) and ripened at 13 °C for six months. Samples were taken from the inner part of the cheese at 0 (fresh cheese before salting), 0.5, 1, 2, 4, and 6 months of ripening and stored at −20 °C until analysis.

2.2. Compositional and Microbiological Analyses of Cheeses

Moisture content and pH of the RO-milk and cheeses were measured using a Mettler-Toledo HR83 moisture analyzer (Mettler-Toledo AG, Greifensee, Switzerland) [25] and a pH meter (Mettler-Toledo Ltd., Leicester, UK), respectively. The pH of the cheese was measured by inserting a glass electrode into the compressed grated cheese samples. Total fat content of the cheese was determined by the method of the Association of Official Analytical Chemists AOAC 933.05 [26] in the grated lyophilized cheese samples.

2.3. Analysis of Caseins

Fractions of CN and their primary degradation products in cheeses were analyzed by Agilent Capillary Electrophoresis (CE, Agilent Technologies, Waldbronn, Germany) according to the method of Ardö and Polychroniadou [27] as described by Taivosalo et al. [28]. CN fractions were identified and labeled based on the results of Otte et al. [29], Miralles et al. [30], Albillos et al. [31], and Heck et al. [32]. 'Valley-to-valley' integration of the peaks was used [30]. The ratio of peak area to the total peak area adjusted to the migration time was used to calculate the relative concentration of caseins in the cheese samples [32].

2.4. Analysis of Free Amino Acids

Composition of the free amino acids (FAA) of RO-cheeses was obtained by UPLC (Acquity UPLC, Waters Corp., Milford, MA, USA) controlled by Waters Empower 2.0 software (Waters Corp., Milford, MA, USA) as described by Taivosalo et al. [28]. The absolute concentrations of AA were calculated using standard curves and expressed as mmol kg^{-1} of cheese for total FAA and relative content as mol% for individual FAA.

2.5. Analysis of Volatiles by SPME-GC-TOF-MS

The extraction of volatile compounds from RO-cheese samples was carried out using the headspace solid-phase microextraction (HS-SPME) method based on Bezerra et al. [33]. Grated cheese was measured (0.1 g) into a 20 mL glass autosampler vial capped with a PTFE/silicone septum. Vials were preincubated at 40 °C for 5 min. A SPME fiber (30/50 µm DVB/Car/PDMS Stableflex, length 2 cm; Supelco, Bellefonte, PA, USA) was used to extract the volatile compounds from the headspace for 20 min under stirring at 40 °C.

Identification of cheese volatiles was performed using a Micromass GCT Premier gas chromatograph system (Waters, Milford, MA, USA) coupled with a CombiPAL autosampler (CTC Analytics AG, Lake Elmo, MN, USA). After the SPME procedure, the volatile compounds were desorbed in splitless mode into a GC injection port equipped with a 0.75 mm internal diameter liner at 250 °C for 10 min. A DB5-MS column (30 m length × 0.25 mm i.d. × 1.0 µm film thickness; J&W Scientific, Folsom, CA, USA) was used with helium as a carrier gas at a flow rate of 1.0 mL min^{-1}. GC conditions were based on the method employed by Lee et al. [34]. The initial oven temperature was 35 °C with a holding time of 3 min. Then, the oven was programmed to ramp-up from 35 °C at a rate of 5 °C min^{-1} to 110 °C, and then from 110 °C at a rate of 10 °C min^{-1} to a final temperature of 240 °C with an additional holding time of 4 min (35 min of total run time). Mass spectra were obtained at ionization energy of 70 eV and a scan speed of 10 scans s^{-1}, with a mass-to-charge ratio scan range of 35 to 350. Three analytical replicates were used for each cheese.

Nontargeted identification of volatile compounds was carried out using the ChromaLynx application (version 4.1; MassLynx software; Waters, Milford, MA, USA) and theoretical calculation of Kovats retention indices (RI). Theoretical RI were calculated using the retention times (RT) of the eluting compounds normalized to the RT of adjacent n-alkanes. Accurate identification of the compounds was verified by comparing theoretical RI to the NIST database (US Department of Commerce, Gaithersburg, MD, USA). The quantities of identified compounds were expressed in peak area units (AU) and as a percentage of a peak area against total ion chromatogram (%TIC).

2.6. Sensory Analysis

Descriptive sensory analysis was performed by a local sensory panel of eight trained assessors. The testing rooms were in compliance with ISO standard [35]. In total, 35 attributes (11 for odor, five for appearance, 14 for flavor, and five for texture) were assessed on a scale of 0–15. Commercial 6-month-old Old Saare (Saaremaa Dairy Factory, Kuressaare, Estonia), made with both mesophilic and thermophilic starters, and 8-month-old Gouda (Valio Eesti AS, Võru, Estonia) cheeses were chosen as references. A complete list of sensory attributes, attribute definitions, and anchor-values of the reference materials on the scale is presented in Table 1.

For sensory analysis, cheese samples were cut into 6 cm × 1 cm × 1 cm pieces. Three pieces of each sample were served to panel members on a white plate. As an exception, appearance attributes were assessed separately from a cross-section of the cheese wheel. Randomized blind-tasting with sequential monadic order of presentation was used. Two replicate assessments were done for each cheese sample. Water was provided as a palate cleanser between the samples.

Table 1. A complete list of sensory attributes, attribute definitions, reference materials, and their anchor-values on the scale.

Sensory Attributes	Attribute Definition	Reference	
		Commercial 6-Month-Old Old Saare Cheese	Commercial 8-Month-old Gouda Cheese
Appearance			
Color	Indicates overall color hue of the sample. The highest score on the scale implies deep orange hue of the sample; the lowest score – off-white hue of the sample	8	15
Hole size	Indicates the average size of holes. The attribute is assessed from the cross-section of a cheese wheel The highest score on the scale implies the presence of large holes in the cross-section; the lowest score—no holes are present in the cross-section of a cheese wheel	0	5
Hole shape	Indicates uniformity and roundness of holes. The attribute is assessed from the cross-section of a cheese wheel. The highest score on the scale implies that the holes (if present) are all round and even; the lowest score—the holes (if present) and misshapen and uneven	0	15
Hole distribution	Indicates the degree of evenness of hole distribution. The attribute is assessed throughout the cross-section of a cheese wheel. The highest score on the scale implies that there is an even distribution of the holes (if present); the lowest score – uneven distribution of the holes (if present)	0	15
Hole merging	Indicates the degree of hole merging (webbing). The attribute is assessed throughout the cross-section of a cheese wheel. The highest score on the scale implies that there is a sever merging of holes; the lowest score—the absence of visible merging	0	0
Odor *			
Intensity	Indicates the strength of the overall perceived odor	8	12
Milk	Indicates overall strength of odor characteristic to milk-based products	8	5
Sour	Indicates overall strength of all sour odors	6	3
Sweet	Indicates overall strength of all sweet odors	5	10
Buttery	Indicates the strength of odor sensation characteristic to butter	5	2
Animalic	Indicates the strength of odor sensation characteristic to musky civet and castoreum [1]	1	0
Sulfur	Indicates the strength of odor sensation characteristic to rotten eggs [1]	0	0
Animal feed	Indicates the strength of odor sensation characteristic to cattle feed yards [1]	0	0
Rancid	Indicates the strength of odor sensation characteristic to products of oxidation [1]	0	0
Yeasty	Indicates the strength of odor sensation characteristic to active yeast [1]	0	0
Metallic	Indicates the strength of odor sensation characteristic to metal or steel [1]	0	0
Flavor *			
Intensity	Indicates the overall strength of perceived flavor (basic taste + retronasal olfaction)	10	13
Sweet	Indicates the strength of overall sweet sensation (basic taste + retronasal olfaction)	12	8
Caramel	Indicates the strength of retronasal olfaction sensation characteristic to caramel, which is formed as a result of cheese maturation	0	5
Sour	Indicates the strength of overall sour sensation (basic taste + retronasal olfaction) characteristic to acids formed as a result of fermentation	4	6

Table 1. *Cont.*

Sensory Attributes	Attribute Definition	Reference: Commercial 6-Month-Old Old Saare Cheese	Reference: Commercial 8-Month-old Gouda Cheese
Bitter	Indicates the strength of bitter taste characteristic to small peptides in cheese (basic taste)	2	3
Salty	Indicates the strength of salty taste characteristic to table salt (basic taste)	5	8
Umami	Indicates the strength of savory taste characteristic to monosodium glutamate (basic taste)	8	6
Milk	Indicates the strength of retronasal olfaction characteristic to milk-based products	8	4
Animalic	Indicates the strength of retronasal olfaction sensation characteristic to musky civet and castoreum [2]	2	0
Sulfur	Indicates the strength of retronasal olfaction sensation characteristic to rotten eggs [2]	0	0
Animal feed	Indicates the strength of retronasal olfaction sensation characteristic to cattle feed yards [2]	0	0
Rancid	Indicates the strength of retronasal olfaction sensation characteristic to products of oxidation [2]	0	0
Yeasty	Indicates the strength of retronasal olfaction sensation characteristic to active yeast [2]	0	0
Metallic	Indicates the strength of sensation in the mouth characteristic to metal or steel [2]	0	0
Texture			
Crumbliness	Indicates the number of particles released when breaking the sample in half. The highest score on the scale implies that no particles were released when breaking the sample in half (the sample does not crumble); the lowest score implies a significant release of particles (the sample crumbles)	3	1
Hardness	Indicates the force required to bite through the sample. The highest score on the scale implies that a lot of force is required to make an initial bite through the sample (the samples is hard); the lowest—barely any force is required to bite through the sample (the samples is soft)	5	10
Rubbery	Indicates the rubbery texture characteristic to squeaky cheeses. The highest score on the scale implies that the sample texture is the least similar to that of squeaky cheeses (the samples is not rubbery); the lowest score on the scale implies that the sample texture is similar to that of squeaky cheeses (the samples is rubbery)	3	7
Adhesiveness	Indicates the amount of sample particles that remain on the teeth after chewing the sample for 5 times. For the adhesiveness assessment, a bite of approx. 1 cm × 1 cm piece should be taken. The highest score on the scale implies that the sample leaves behind a significant residue on the teeth after chewing and swallowing (the samples is adhesive); the lowest—no residue is left behind on the teeth after chewing and swallowing (the sample is not adhesive)	2	8

* The highest score on the scale implies very intense sensation; the lowest score – no sensation.[1] Possible off-odor; [2] possible off-flavor.

3. Results and Discussion

A cheese, produced in this study from RO-milk concentrated 1.9-fold on a dry matter basis, was characterized as hard cheese, with the calculated moisture in nonfat substance (MNFS) content of 45.5% (w/w) and fat content of 37.1% (w/w) as determined in 0.5-month-old cheese and pH ranging from 5.10 to 5.23 during ripening for six months.

3.1. Proteolysis During RO-Cheese Ripening

Figure 1 shows the electropherograms of the intact CN fractions (α_{s1}-CN (8P and 9P), α_{s2}-CN (11P, 12P, and nP), β-CN (genetic variants A^1, A^2, and B), and para-κ-CN), and their primary hydrolysis products (α_{s1}-I-CN (8P and 9P), γ_1-CN (A^1 and A^2), and γ_2-CN and γ_3-CN) identified in RO-cheese during ripening.

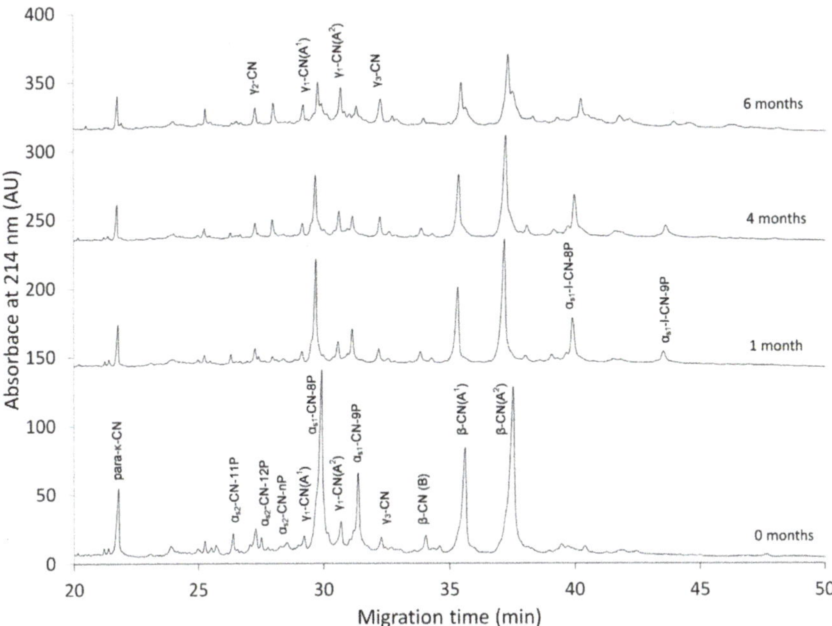

Figure 1. The electropherograms of RO-cheese obtained by CE at 0, 1, 4, and 6 months of ripening. CN: casein; para-κ-CN: κ-CN f1–105; γ_1-CN: β-CN f29–109; γ_2-CN: β-CN f106–209; γ_3-CN: β-CN f107–209; α_{s1}-I-CN-8P: α_{s1}-CN f24–199; α_{s1}-I-CN-9P: α_{s1}-CN f24–199 9P; A^1, A^2, and B: genetic variants of β-CN; 11P, 12P, 9P, 8P, and nP: phosphorylation states of caseins.

Intact α_{s1}-CN was subjected to more rapid degradation than β-CN: approximately 60% of α_{s1}-CN breakdown was observed within the first month and 70% was hydrolyzed by the end of the sixth month of ageing (Figures 1 and 2a). The peaks corresponding to β-CN also showed obvious changes throughout ripening, but still ~55% of the initial β-CN remained intact by the end of the ripening period. The rate of the hydrolysis of the initial CN fraction in RO-cheese was the highest during the first months of ripening (Figure 2).

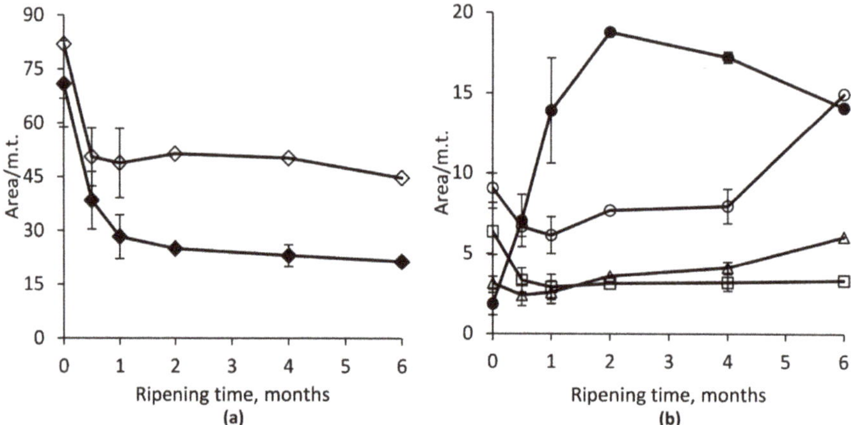

Figure 2. Change of the content (normalized peak area) of main intact CN (**a**) and their primary degradation products (**b**) during RO-cheese ripening. α_{s1}-CN (♦): sum of α_{s1}-CN-8P and α_{s1}-CN-9P; β-CN (◊): sum of β-CN(A^1), β-CN(A^2), and β-CN(B); α_{s1}-I-CN (●): sum of α_{s1}-I-CN-8P and α_{s1}-I-CN-9P; γ_1-CN (○): sum of γ_1-CN(A^1) and γ_1-CN(A^2); (□): γ_2-CN; (Δ): γ_3-CN; m.t., migration time.

Due to the lack of published data on RO-cheeses and the large diversity of the manufacturing parameters of traditional cheeses, a direct comparison of the ripening characteristics of our RO-cheese to those of other cheeses was fairly complicated. Nevertheless, it was instantly evident that the course of primary proteolysis in RO-cheese was comparable with those of traditional cheeses. In traditional cheese varieties with similar manufacturing technology, primary proteolysis is characterized by a rapid breakdown of intact caseins to high molecular weight peptides by an activity of both the rennet (chymosin) retained in the curd and the milk native proteinase plasmin. The addition of rennet based on the initial amount of milk used for RO-milk production resulted in an adequate degree of primary degradation of α_{s1}-CN in our RO-cheese (Figures 1 and 2a), which is consistent with results on proteolysis in RO-Cheddar cheese [23]. Quite similar results on the breakdown of α_{s1}-CN have been reported in a traditional hard cheese Old Saare, with manufacturing and ripening conditions very similar to our RO-cheese (a high curd scalding temperature and both mesophilic and thermophilic LAB as starters), where the same share—~70% of α_{s1}-CN—was hydrolyzed during six months of ripening [28]. In 26-week-ripened traditional extra-hard Västerbottenost cheese, manufactured with high scalding temperature and mesophilic starter, the initial peaks of both α_{s1}- and β-CN had almost completely disappeared according to the CE profiles [36]. Proteolysis in a typical Gouda-type cheese is determined by an ~70–80% decrease in α_{s1}-CN content, mainly by the action of chymosin, already within the first month of ripening [37–39], which is consistent with our results for α_{s1}-CN degradation; during the production of RO-cheese, the curd scalding temperature was higher (48 °C) compared to Gouda-type cheeses, which could have caused a partial inactivation of chymosin that affected the degradation of α_{s1}-CN [40]. Indeed, a rather high percentage of α_{s1}-CN breakdown determined in our cheese can be associated with a rather low pH of the cheese environment (in the range of 5.10 to 5.23 during the ripening period) being more favorable for the activity of chymosin as well as also the reversible and incomplete thermal inactivation of chymosin after cooking the cheese curds at 48 °C [40,41].

The substantial contribution of plasmin to the primary proteolysis in cheeses with a high curd cooking temperature is evident [42]. Nevertheless, conversely to α_{s1}-CN degradation, significantly less β-CN was hydrolyzed in RO-cheese (45% of intact β-CN hydrolyzed by the end of ripening) compared to the levels reported in Old Saare and even in Gouda-type cheeses (78% and 60–50% of β-CN hydrolyzed by six months of ripening, respectively) [28,37–39]. The greater activity of plasmin on β-CN could be expected in RO-cheese, as high cooking temperature inactivates inhibitors of plasminogen activators, leading to the conversion of plasminogen to the active plasmin [43]. Both the higher curd scalding temperature (52 °C) and higher pH during ripening (5.3–5.5) of Old Saare [28], as well as in Västerbottenost cheeses (5.3–5.6) [36], than that of RO-cheese could have exerted an impact on more intensive primary hydrolysis of β-CN in those cheeses [43]. Considerable lower pH values in RO-cheese were evidently decisive for the reduced activity of the alkaline proteinase, plasmin [43], which resulted in impaired hydrolysis of β-CN in RO-cheese.

Large peptides obtained from the primary hydrolysis serve as substrates for subsequent generation of shorter peptides and FAA by complex action of different proteolytic enzymes from LAB [39]. Figures 1 and 2b show that the production rate of chymosin-derived peptide α_{s1}-I-CN was the highest during the first two months, while during the next four months of ageing, the degradation rate of that peptide became higher than the production. At the beginning of the ripening, the content of plasmin-derived peptides γ-CNs decreased moderately during the first month, suggesting that they could have been more rapidly degraded further to shorter peptides by the LAB enzymes. The γ-CNs accumulated during further months of aging, showing an increase in the production rate in the period between four and six months (Figures 1 and 2b).

The final step of proteolysis is the release of FAA, those enzymatic and chemical conversions to volatile compounds lead to the development of the characteristic cheese flavor [39,44]. The total and relative (mol%) content of FAA released during six months of RO-cheese ripening is shown in Figures 3 and 4, respectively.

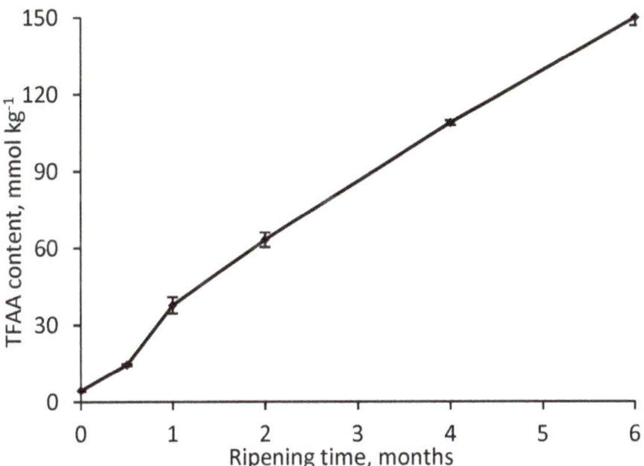

Figure 3. The change in total free amino acids (TFAA) content during RO-cheese ripening.

The total content of the FAA increased from 4.3 ± 0.3 to 149.9 ± 3.2 mmol kg^{-1} of cheese during maturation. The use of a thermophilic starter in cheese production could have considerably increased the total FAA content [45,46] in RO-cheese. Nevertheless, the level of total FAA in 6-month-old RO-cheese was three-fold lower than that in hard Old Saare cheese (450 mmol kg^{-1} of cheese) [28], it is also made with both mesophilic and thermophilic starters but shows intensive degradation of β-CN, unlike the RO-cheese. Moreover, the total FAA content in RO-cheese was somewhat lower than that in 6-month-old extra-hard Västerbottenost (150–260 mmol kg^{-1}) and semihard Herrgård (190 mmol kg^{-1}) cheeses manufactured only with mesophilic starters [36,46]. This clearly reveals that the lower total FAA content in RO-cheese can be attributed to the lower degree of hydrolysis of one of the main casein—β-CN—on the first stage of proteolysis mediated by plasmin activity.

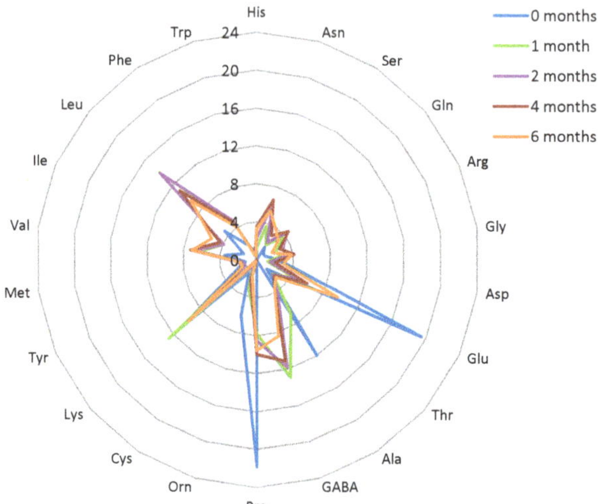

Figure 4. Radar diagram showing the relative content of individual free amino acids (FAA) during six months of RO-cheese ripening (mol%). Results presented are the means of two measurements. Amino acids: GABA, γ-aminobutyric acid; Ala, alanine; Thr, threonine; Glu, glutamate; Asp, aspartate; Gly, glycine; Arg, arginine; Gln, glutamine; Ser, serine; Asn, asparagine; His, histidine; Trp, tryptophan; Phe, phenylalanine; Leu, leucine; Ile, isoleucine; Val, valine; Met, methionine; Tyr, tyrosine; Lys, lysine; Cys, cysteine; Orn, ornithine; and Pro, proline.

The quantitatively dominating amino acids in the ripened 6-month-old RO-cheese cheese were Lys, Pro, Glu, Leu, and γ-aminobutyric acid (GABA) (Figure 4). The first four were found at very similar concentrations (14.2–14.8 mmol kg^{-1} of cheese). It is noteworthy that the same four AA have been shown to be among the quantitatively dominating ones in matured Old Saare cheese [28]. A large amount of Pro (14.4 mmol kg^{-1} in ripened RO-cheese) obviously produced by peptidases of *Lb. helveticus*, may introduce the sweet note in the cheese flavor profile [47]. Amino acid Val, reported to be within the quantitatively dominating amino acids in ripened Old Saare cheese [28], was determined within the dominating ones in the middle of the RO-cheese ripening process (2–4 months), and may contribute to the bitter and sweet flavors of cheese [47]. High levels of Lys and Leu could have contributed to the bitter note [47] in the taste of ripened RO-cheese. GABA was determined in large amounts from the first month of ripening. GABA is not originally present in caseins, but several LABs, including *St. thermophilus* and *Lb. helveticus*, have been shown to exert a GABA-producing ability through the decarboxylation of Glu [47,48].

3.2. Volatile Compounds

A total of 42 volatile compounds were identified in RO-cheese: nine alcohols, six ketones, seven acids, seven esters, six aldehydes, four aromatic compounds, and three of other groups. Table 2 shows the volatile compounds grouped by classes and the changes in content throughout the ripening process. Volatile acids and alcohols were the main volatiles identified in RO-cheese (1.5–46.3 and 0.0–10.4%TIC, respectively). Esters and ketones comprised the smaller share (0.0–6.7 and 0.0–2.9%TIC, respectively, except for the ketones at the initial point of ripening), while aldehydes and aromatic compounds were among the minor ones (0.0–0.3 and 0.0–0.4%TIC, respectively). The identified volatile compounds showed a dynamic behavior during cheese maturation; however, the total amounts within most detected chemical groups (except for the ketones and aromatic compounds) increased by the end of ripening (Figure 5).

Table 2. Volatile compounds identified in RO-cheese during ripening (AU × 10^4). Results are the means of three GC-MS measurements. AU, arbitrary units (peak area).

Compound	RT	RI, Exp	RI, Theor [1]	Odor Description [2]	Ripening Time, Months					
					0	0.5	1	2	4	6
Alcohols										
Isopropyl Alcohol	2.15	510	515	Woody, musty	–	t	–	–	t	0.10
2-Butanol, 3-methyl	7.92	715	700	Musty, vegetable, cheesy	1.34	–	–	–	–	–
3-Buten-1-ol, 3-methyl-	8.33	725	720	Fruity	–	0.04	–	–	–	–
1-Butanol, 2-methyl-	8.34	726	740	Fruity, whiskey	–	–	–	–	–	0.04
1-Butanol, 3-methyl-	8.80	738	750	Fruity, banana, cognac	–	–	–	–	0.09	0.18
2,3-Butanediol	10.34	778	779	Creamy, buttery	0.15	1.72	–	1.73	t	0.86
1,3-Butanediol	10.96	794	789	Odorless	–	–	–	0.32	2.57	2.47
1-Hexanol, 2-ethyl	18.93	1009	1025	Fruity, floral, fatty	0.02	0.03	–	–	–	0.03
1-Undecanol	26.03	1358	1370	Soapy, citrus	–	0.04	–	–	–	–
Total					1.51	1.85	0.00	2.05	2.66	3.69
Aldehydes										
Butanal, 3-methyl-	5.49	640	652	Fruity, cocoa, nutty	–	–	–	–	–	0.03
Butanal, 2-methyl-	5.78	650	664	Musty, nutty, fermented	–	–	–	–	–	0.06
Octanal	18.49	995	1000	Citrus, orange peel, waxy	–	0.06	–	–	–	–
Nonanal	20.97	1085	1099	Citrus, green, cucumber	0.01	0.03	0.01	–	t	–
Decanal	23.08	1183	1188	Citrus, orange peel, waxy	0.02	–	–	–	–	–
Undecanal	25.07	1295	1310	Soapy, citrus	–	0.02	–	–	–	–
Total					0.03	0.12	0.01	0.00	0.00	0.09
Ketones										
Acetone	2.12	509	509	Solvent, apple, pear	0.24	0.30	–	–	–	–
2,3-Butanedione	3.68	577	574	Buttery, creamy, milky	0.17	–	–	0.17	–	0.94
2-Pentanone	7.26	697	697	Fruity, banana, fermented	7.79	–	–	–	–	–
2-Butanone, 3-hydroxy	7.42	711	706	Creamy, dairy, butter	–	–	t	–	0.13	0
2-Heptanone	14.00	873	887	Cheesy, fruity, banana	0.04	–	–	–	0.01	0.06
2-Nonanone	20.62	1072	1090	Fruity, dairy, soapy	t	–	–	–	–	–
Total					8.23	0.30	0.00	0.17	0.15	1.01
Acids										
Acetic acid	5.25	633	640	Pungent	0.71	t	0.47	4.05	14.02	12.62
Propanoic acid	7.11	692	700	Pungent, dairy	–	–	–	–	0.00	–
Propanoic acid, 2-methyl-	9.15	747	758	Acidic, cheesy, dairy	–	–	–	–	0.05	0.04
Butanoic acid	10.14	773	790	Sharp, cheesy	0.01	0.00	003	0.50	1.17	t
Butanoic acid, 3-methyl	12.35	830	848	Cheesy, dairy, fermented, berry	–	–	–	0.00	0.38	0.21
Butanoic acid, 2-methyl	12.74	840	846	Fruity, dairy, cheesy	–	–	–	–	0.05	0.09
Hexanoic acid	17.34	963	990	Fatty, cheesy	–	t	–	0.05	0.11	–
Total					0.72	t	0.50	4.60	15.80	12.95

Table 2. *Cont.*

Compound	RT	RI, Exp	RI, Theor [1]	Odor Description [2]	Ripening Time, Months					
					0	0.5	1	2	4	6
Esters										
Ethyl Acetate	4.44	607	610	Ethereal, fruity, grape, cherry	–	t	0.01	–	t	t
Butanoic acid, ethyl ester	10.91	793	798	Fruity, sweet, apple	–	0.07	0.04	t	0.55	2.03
Butanoic acid, butyl ester	18.09	984	994	Fruity, banana, pineapple	–	–	–	0.08	–	0.02
Hexanoic acid, ethyl ester	18.23	988	996	Fruity, banana, pineapple, green	–	–	0.06	0.13	0.06	0.23
Octanoic acid, ethyl ester	23.00	1179	1190	Fruity, pineapple, musty	–	–	–	–	t	0.07
Decanoic acid, ethyl ester	26.27	1373	1381	Fruity, apple	–	–	–	–	t	0.02
Dodecanoic acid, ethyl ester	28.97	1572	1581	Floral, creamy, dairy, soapy	–		–	–	–	0.01
Total					–	0.07	0.11	0.20	0.61	2.37
Aromatics										
Toluene	9.43	770	774	Sweet, pungent, caramel	t	0.02	0.01	–	0.01	–
Benzaldehyde	16.87	950	955	Almond, cherry	0.06	0.06	0.02	–	0.00	–
Acetophenone	20.17	1055	1062	Almond, cherry, fruity, floral	t	0.02	0.04	0.02	0.02	–
Indole	24.84	1282	1292	Animal, fecal	t	–	–	–	–	–
Total					0.06		0.10	0.02	0.03	–
Others										
n-Butyl ether	14.05	883	888	Ethereal	–	t	–	0.07	0.01	0.05
Dimethyl sulfone	14.94	897	918	Sulfur	t	0.06	–	–	0.03	0.03
Dimethyl sulfide	2.45	523	530	Sulfur, onion, cabbage	t	0.02	t	–	–	–
Total					t	0.08	t	0.07	0.05	0.09

[1] NIST database (US Department of Commerce, Gaithersburg, MD, USA). [2] The Good Scents Company Information System (Oak Creek, WI, USA). t (traces), in nonquantifiable amounts; –, not detected.

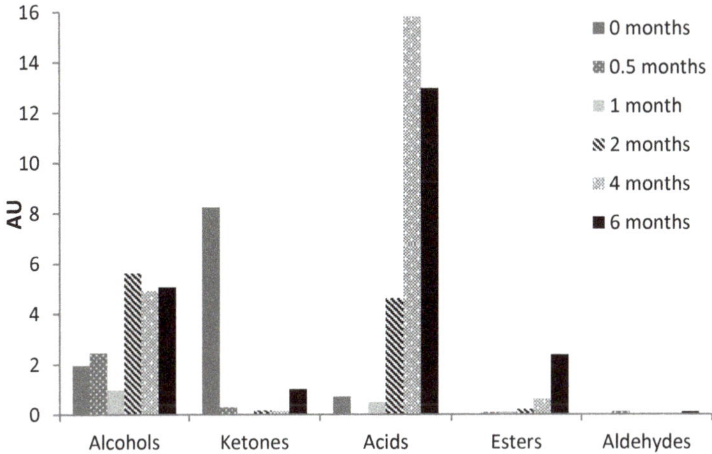

Figure 5. Changes in the content of the main chemical groups of the volatile compounds (AU × 10^4) identified during RO-cheese ripening. AU: arbitrary units (peak area).

3.2.1. Carboxylic Acids

The total amount of volatile carboxylic acids increased during the ripening of RO-cheese for six months, with a slightly higher content in 4-month-old than in 6-month-old cheese (Figure 5). The relative content of carboxylic acids was the highest in the 4-month-old cheese (46.3%TIC), but also retained high levels toward the end of the ripening period. Acetic acid was the most abundant compound with a very high relative content (1.3–41.0%TIC) during all stages of cheese maturation. The content of acetic acid increased greatly after one month of ripening (Table 2). Acetic and propanoic acids may have a microbial origin and could have been formed as a product of lactose metabolism [44]. Hexanoic acid was present in relatively low levels, exhibiting an ascending tendency solely in the middle of RO-cheese ripening until disappearing after the fourth month of ripening. Butanoic acid was present in comparable amount at the beginning of ripening and showed a substantial increase in amounts up to the fourth month of maturation and disappeared thereafter. Short-chain fatty acids and butanoic and hexanoic acids could be produced as a result of the enzymatic hydrolysis of triglycerides, which is also one of the biochemical pathways essential for cheese flavor development [35]. It is likely that the linear-chain fatty acids could have been transformed into esters [49] since their content was higher at the end of RO-cheese ripening (Figure 5). The main branched-chain fatty acids in RO-cheese—2- and 3-methylbutanoic acid—were present in substantial amounts after four months of ripening (Table 2) and could provide sweaty, sour, fruity, and buttery flavor notes [47,49]. Methylated acids as well as methylated aldehydes and methylated alcohols have a proteolytic origin and, obviously, are produced by the catabolism of branched-chain amino acids (Leu, Ile, and Val) by aminotransferases [47,49]. The prompt decreased in the relative content of Leu in RO-cheese from the fourth month of ripening could be associated with the higher content of 3-methyl-butanoic acid, 3-methyl-1-butanol, and 3-methyl-butanal in ripened 4- and 6-month-old RO-cheeses. In addition, the relative content of Val did not increase much after the second month of ripening, which correlated to the appearance of 2-methylpropanoic acid. Volatiles 2-methylbutanoic acid, 2-methyl-1-butanol, and 2-methyl-butanal originate from Ile, but we could not relate the occurrence of these components in the ripened RO-cheeses to the content of Ile, as it still steadily increased during the RO-cheese maturation. All volatile carboxylic acids identified in RO-cheese (except for propanoic acid) have been previously reported as aroma-active in a wide range of young, medium, and aged traditional Gouda-type cheeses by SPME-GC-Olfactometry [50]. In addition, the same acids (except for acetic and propanoic acid) have been detected in 6-week-old Gouda-type cheeses, while aged cheeses have not shown them at all [51].

Acetic, butanoic, and hexanoic acids, as determined by SPME-GC-Olfactometry, have been found to contribute greatly to the characteristic cheesy sharp and mild to strong savory aroma of typical hard Parmesan and Grana Padano cheeses [52]. Along with the mentioned carboxylic acids, octanoic and decanoic acids were among the abundant aroma compounds in Parmigiano-Reggiano of different ages reported by Bellesia et al. [53].

3.2.2. Alcohols

Alcohols were the second major class of volatile compounds identified in 0-month-old cheese (6.5%TIC) and in cheeses after the second month (up to 10.4%TIC) of ripening (Table 2). The total alcohol content showed an increasing trend throughout cheese maturation up to the second month of ripening with a more rapid change of content after the first month. The total content of alcohols was more or less similar during further RO-cheese ripening (Figure 5). Among the alcohols, 3-methyl-2-butanol was abundant in fresh cheese, whereas in the middle and the later stages of ripening dominated 2,3- and 1,3-butanediol, respectively (Table 2). 2,3-Butanediol could be formed during the citrate or Asp metabolism from 2,3-butanedione (diacetyl) [44], and has been previously reported to be among the important flavor compounds in 0.5- and 4-month-old Gouda cheese [51]. Branched-chain alcohols—2-methyl-1-butanol and 3-methyl-1-butanol—and their corresponding methylated aldehydes—2- and 3-methylbutanal—identified in RO-cheese have been previously detected among the aroma-active compounds in extra-hard Västerbottenost [36], as well as in young and aged Gouda-type cheeses [50,51]. The levels of these compounds were reported to be higher in the matured cheeses, which is consistent with our results obtained for RO-cheese, where these alcohols were detected after four months of ripening. 3-Methyl-2-butanol, 3-methyl-1-butanol, and 2,3- and 1,3-butanediol were detected at trace amounts or among the less abundant alcohols in Parmigiano-Reggiano [53].

3.2.3. Esters

Esters were found in relatively small amounts in the volatile fraction of RO-cheese. This chemical group was presented at a low level from the second week until the second month of ripening and then increased at the later stages of cheese ripening (Figure 5). Butanoic acid ethyl ester and hexanoic acid ethyl ester were dominant among the esters during the entire period of ripening (Table 2). The former was present especially in high amounts in the 4- and 6-month-old cheese and composed 1.6 and 5.7%TIC, respectively. Octanoic acid, decanoic acid, and dodecanoic acid ethyl esters were found only in the 6-month-old cheese. Ethyl esters originate from the enzymatic or chemical esterification of the fatty acids and characterize a cheese by sweet and fruity notes [44]. Butanoic acid ethyl ester and hexanoic acid ethyl ester have been shown to have the highest contribution to the cheese flavor within the esters in young and matured Gouda [50] and hard Parmesan-type and Grana Padano [52,53]. Other ethyl esters identified in RO-cheese have also been quantified, but only in matured Gouda-type cheeses [51], which is in agreement with our results.

3.2.4. Ketones

Ketones were the most abundant class in fresh cheese (35.4%TIC), showing very high amounts compared to the other chemical groups of volatile compounds identified in RO-cheese, mainly because of 2-pentanone (Table 2 and Figure 5). The level of ketones decreased promptly by the second week of ripening. Ketones were found in low levels at the middle stages of maturation. A substantial amount of ketones was observed again at the end of the ripening process in the 6-month-old RO-cheese, mainly because of the increase in content of 2,3-butanedione, which was observed at all stages of ripening. 2,3-butanedione (diacetyl) is one of the products of citrate or Asp metabolism with a sweet buttery aroma [44] and has been demonstrated to be an important aroma-active compound in hard Parmesan cheese [52], considered to be characteristic to Gouda cheeses of different ages [50]. Methyl ketones (2-pentanone, 2-heptanone, and 2-nonanone) can be produced from fatty acids through β-decarboxylation and may be transformed to secondary alcohols [44]. Methyl ketones have been identified as important constituents in blue-cheese aroma [52], although small amounts of those have also been observed in some Gouda [50,51] and Parmigiano-Reggiano of different ages [53]. Moreover, methyl ketones have been found in the fraction of volatile ketones in Västerbottenost cheese with the highest abundance of 2-pentanone [36].

3.2.5. Aldehydes

Aldehydes were present among the volatiles of RO-cheese in very small amounts with a fluctuating behavior during ripening (Table 2). The total content of aldehydes was the highest both in the 0.5- and 6-month-old cheese (0.12 and 0.09%TIC) (Figure 5). Only 2- and 3-methylbutanal were detected in matured 6-month-old cheese, whereas cheeses up to one month of ripening contained the linear-chain aldehydes octanal, nonanal, decanal, and undecanal. Aldehydes are minor volatile components present at low levels in cheese because they are rapidly converted to alcohols or corresponding acids [44]. Aldehydes 2-methyl-butanal and 3-methyl-butanal are the products of the catabolism of branched-chain amino acids Ile and Leu, and have been shown to give the malty, fruity, cocoa, and nutty flavors to cheese [47,49]. The decreasing relative content of Leu in RO-cheese after the fourth month of ripening (Figure 3) can be related to the appearance of low levels of these aldehydes due to their further rapid transformation into the corresponding alcohols 2- and 3-methyl-1-butanol or 2- and 3-methylbutanoic acids (Table 2). The above-mentioned aldehydes have been found among the strong aroma-active compounds at higher concentrations in matured 9- and 10-month-old Gouda than in younger cheeses [50]. In addition, 2-methyl-butanal and 3-methyl-butanal have been detected in Västerbottenost [36] and in some Parmigiano-Reggiano cheeses [53].

3.2.6. Aromatics and Other Compounds

Among the aromatic compounds benzaldehyde, toluene, and acetophenone were found in RO-cheeses up to the fourth month of ripening with a larger share of benzaldehyde in the two first ripening points. Benzaldehyde can be produced from the aromatic amino acid Phe via the α-oxidation of phenyl acetaldehyde and have been shown to give notes of bitter almond to aged Gouda [50,51], Parmigiano-Reggiano [53], and Västerbottenost cheeses [36]. Acetophenone has been found in aged Gouda [50]. Sulfur compounds derive from the amino acid Met and are essential components in many cheeses, giving the boiled cabbage and potatoes, garlic, and egg flavors [47].

3.3. *Sensory Properties*

Within sensory perception, an appearance modality revealed the most obvious changes during RO-cheese maturation. The 6-month-old cheese was considerably darker and richer in color than the young RO-cheeses (Figure 6). The size of holes within the RO-cheese matrix grew rapidly during the first month of ripening and then remained relatively the same throughout maturation. The distribution of holes within the cheese matrix became more uniform after the first month of ripening, and some partial merging of the holes was noticed at all ripening stages. The rubberiness decreased, and crumbliness increased dramatically during ripening. In matured 6-month-old RO-cheese, small, white crystals were observed and perceived in the interior of the cheese and on the surface of the holes. The crystals are commonly formed due to the crystallization of amino acids, e.g., Tyr, or calcium lactate, when lactobacilli-containing (including *Lactobacillus helveticus*) starters are used, and have been shown to occur after prolonged maturation of Gouda-type, Cheddar, and Parmesan cheeses [38,54].

Figure 7 shows the principal component analysis (PCA) carried out on the scores of RO-cheese odor and taste evaluation. The first two principal components (PC) explained 81.49% of the variability (PC1: 58.72%; PC2: 22.77%). The odor and taste attributes were related more to the younger, up to 2-month-old cheeses located on the negative axis of PC1, whereas on the positive axis, the attributes received high scores for the matured 4- and 6-month-old RO-cheeses. Young cheeses were characterized by high scores of milky odor and taste and buttery odor, which diminished with maturing time. With maturation, the cheese became more intense in overall intensity (14 out of 15 for odor; 12 for taste), sweetness (8 for odor; 10 for taste), saltiness (seven), and umami taste (seven), and gained low scores for bitterness (three) and caramel taste (two). A slight yeasty odor and flavor (both scored 0.5) were noted in the ripened 6-month-old cheese. These results of the sensory evaluation of odor and taste of ripened 6-month RO-cheese were comparable with those reported for aged 9- and 12-month-old traditional Gouda, where cheeses have been characterized by sweet, salty, and umami tastes and low intensities of caramel and fruity notes [50]. However, an opposite trend in the development of the sweet attribute in cheese maturation, compared to the RO-cheese, has also been observed in Gouda cheese [51].

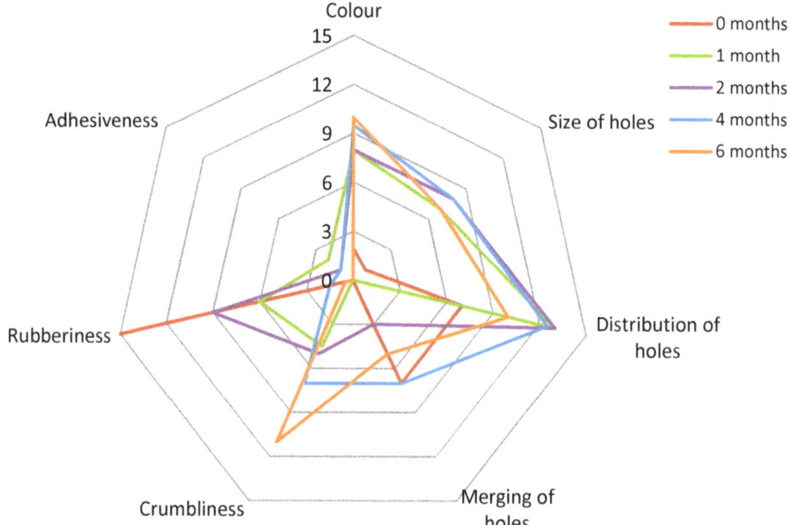

Figure 6. Radar diagram of the appearance and texture attributes of RO-cheese during ripening.

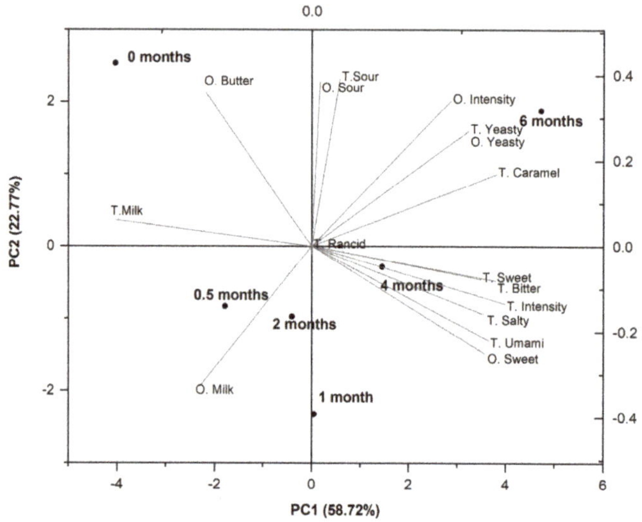

Figure 7. Principal component analysis (PCA) biplot of odor (O) and taste (T) attributes of RO-cheese during ripening. Cheeses are indicated by ripening time. PC: principal component.

No defects in descriptive texture and no off-flavor formation were detected during the descriptive sensory analysis of RO-cheese.

4. Conclusions

In the present study, the hard high cooked cheese was produced in a pilot plant from 1.9-fold concentrated RO-milk. The RO-cheese produced and evaluated in this study was considered to be of satisfactory quality based on sensory testing by a panel of trained assessors. The evaluation of the primary proteolysis, formation of FAA and volatile compounds proved that the patterns of biological processes essential for cheese maturation which took place in RO-cheese are essentially similar to those in traditional Gouda-type and other hard-type cheeses.

Author Contributions: Conceptualization, A.S., N.P., and I.S.; Methodology, T.K., N.P., and I.S.; Investigation, A.S., N.P., and I.S.; Instrumental and Data Analyses, I.S., N.P., J.R., and A.T.; Writing—Original Draft Preparation, A.T.; Writing—Review and Editing, A.T., T.K., J.R., N.P., and I.S.; Visualization, A.T. and J.R.; Project Administration, N.P., A.S., and R.V.; Funding Acquisition, R.V.

Funding: The research was funded by Enterprise Estonia project number EU48667.

Acknowledgments: We thank Elpa OÜ for providing the concentrated milk. We thank Signe Adamberg for critically revising the manuscript and helpful suggestions.

Conflicts of Interest: The authors declare no conflicts of interest.

References

1. Gosalvitr, P.; Franca, R.C.; Smith, R.; Azapagic, A. Energy demand and carbon footprint of cheddar cheese with energy recovery from cheese whey. *Energy Procedia* **2019**, *161*, 10–16. [CrossRef]
2. Mistry, V.V.; Maubois, J.-L. Application of membrane separation technology to cheese production. In *Cheese: Chemistry, Physics and Microbiology*; McSweeney, P.L.H., Fox, P.F., Cotter, P.D., Everett, D.W., Eds.; Elsevier: Oxford, UK, 2017; pp. 677–697.
3. Guinee, T.P.; O'Kennedy, B.T.; Kelly, P.M. Effect of milk protein standardization using different methods on the composition and yields of Cheddar cheese. *J. Dairy Sci.* **2006**, *89*, 468–482. [CrossRef]
4. Lipnizki, F. Cross-flow membrane applications in the food industry. *Membr. Technol.* **2010**, *3*, 1–24.
5. Delgado, D.; Salazar, G.; García, M. Sequential optimisation of yield and sensory quality of semi-hard cheese manufactured from a mixture of ultrafiltered ewes' and cows' milk. *Int. Dairy J.* **2013**, *32*, 89–98. [CrossRef]
6. Karami, M.; Ehsani, M.R.; Mousavi, S.M.; Rezaei, K.; Safari, M. Changes in the rheological properties of Iranian UF-Feta cheese during ripening. *Food Chem.* **2009**, *112*, 539–544. [CrossRef]
7. Karami, M.; Reza Ehsani, M.; Ebrahimzadeh Mousavi, M.; Rezaei, K.; Safari, M. Microstructural changes in fat during the ripening of Iranian ultrafiltered Feta cheese. *J. Dairy Sci.* **2008**, *91*, 4147–4154. [CrossRef] [PubMed]
8. Karami, M. Enhancing the lipolysis of feta-type cheese made from ultrafiltered cow's milk. *LWT-Food Sci. Technol.* **2017**, *80*, 386–393. [CrossRef]
9. Daufin, G.; Escudier, J.P.; Carrère, H.; Bérot, S.; Fillaudeau, L.; Decloux, M. Recent and emerging applications of membrane processes in the food and dairy industry. *Food Bioprod. Process. Trans. Inst. Chem. Eng. Part C* **2001**, *79*, 89–102. [CrossRef]
10. Brandsma, R.L.; Rizvi, S.S.H. Depletion of whey proteins and calcium by microfiltration of acidified skim milk prior to cheese making. *J. Dairy Sci.* **1999**, *82*, 2063–2069. [CrossRef]
11. Neocleous, M.; Barbano, D.M.; Rudan, M.A. Impact of low concentration factor microfiltration on the composition and aging of Cheddar cheese. *J. Dairy Sci.* **2002**, *85*, 2425–2437. [CrossRef]
12. Amelia, I.; Drake, M.; Nelson, B.; Barbano, D.M. A new method for the production of low-fat Cheddar cheese. *J. Dairy Sci.* **2013**, *96*, 4870–4884. [CrossRef] [PubMed]
13. Schreier, K.; Schafroth, K.; Thomet, A. Application of cross-flow microfiltration to semi-hard cheese production from milk retentates. *Desalination* **2010**, *250*, 1091–1094. [CrossRef]
14. Heino, A.; Uusi-Rauva, J.; Outinen, M. Pre-treatment methods of Edam cheese milk. Effect on cheese yield and quality. *LWT-Food Sci. Technol.* **2010**, *43*, 640–646. [CrossRef]

15. Henning, D.R.; Baer, R.J.; Hassan, A.N.; Dave, R. Major advances in concentrated and dry milk products, cheese, and milk fat-based spreads. *J. Dairy Sci.* **2006**, *89*, 1179–1188. [CrossRef]
16. Kumar, P.; Sharma, N.; Ranjan, R.; Kumar, S.; Bhat, Z.F.; Jeong, D.K. Perspective of membrane technology in dairy industry: A review. *Asian-Australasian J. Anim. Sci.* **2013**, *26*, 1347–1358. [CrossRef] [PubMed]
17. Yorgun, M.S.; Balcioglu, I.A.; Saygin, O. Performance comparison of ultrafiltration, nanofiltration and reverse osmosis on whey treatment. *Desalination* **2008**, *229*, 204–216. [CrossRef]
18. Marx, M.; Bernauer, S.; Kulozik, U. Manufacturing of reverse osmosis whey concentrates with extended shelf life and high protein nativity. *Int. Dairy J.* **2018**, *86*, 57–64. [CrossRef]
19. Balde, A.; Aïder, M. Effect of cryoconcentration, reverse osmosis and vacuum evaporation as concentration step of skim milk prior to drying on the powder properties. *Powder Technol.* **2017**, *319*, 463–471. [CrossRef]
20. Voutsinas, L.P.; Katsiari, M.C.; Pappas, C.P.; Mallatou, H. Production of yoghurt from sheep ' s milk which had been concentrated by reverse osmosis and stored frozen. 1. Physicochemical, microbiological and physical stability characteristics of concentrates. *Food Res. Int.* **1996**, *29*, 403–409. [CrossRef]
21. Agbevavi, T.; Rouleau, D.; Mayer, R. Production and quality of Cheddar cheese manufactured from whole milk concentrated by reverse osmosis. *J. Food Sci.* **1983**, *48*, 642–643. [CrossRef]
22. Barbano, D.M.; Bynum, D.G. Whole milk reverse osmosis retentates for Cheddar cheese manufacture: cheese composition and yield. *J. Dairy Sci.* **1984**, *67*, 2839–2849. [CrossRef]
23. Bynum, D.G.; Barbano, D.M. Whole milk reverse osmosis retentates for Cheddar cheese manufacture: chemical changes during aging. *J. Dairy Sci.* **1985**, *68*, 1–10. [CrossRef]
24. Hydamakai, A.W.; Wilbey, R.A.; Lewis, M.J. Manufacture of direct acidified cheese from ultrafiltration and reverse osmosis retentates. *Int. J. Dairy Technol.* **2000**, *53*, 120–124. [CrossRef]
25. ISO (International Organisation for Standardisation). *Cheese and Processed Cheese—Determination of the Total Solids Content. Multiple*; ISO 5534. 2004; American National Standards Institute (ANSI): Washington, DC, USA, 2007.
26. Association of Official Analytical Chemists—AOAC. Official method 933.05. Fat in cheese. In *Official Methods of Analysis*, 19th ed.; AOAC International: Gaithersburg, MD, USA, 2012.
27. Ardö, Y.; Polychroniadou, A. Analysis of caseins. In *Laboratory Manual for Chemical Analysis of Cheese*; Office for Official Publications of the European Communities: Luxembourg, 1999; pp. 50–53.
28. Taivosalo, A.; Kriščiunaite, T.; Seiman, A.; Part, N.; Stulova, I.; Vilu, R. Comprehensive analysis of proteolysis during 8 months of ripening of high-cooked Old Saare cheese. *J. Dairy Sci.* **2018**, *101*, 944–967. [CrossRef] [PubMed]
29. Otte, J.; Zakora, M.; Kristiansen, K.R.; Qvist, K.B. Analysis of bovine caseins and primary hydrolysis products in cheese by capillary zone electrophoresis. *Lait* **1997**, *77*, 241–257. [CrossRef]
30. Miralles, B.; Ramos, M.; Amigo, L. Influence of proteolysis of milk on the whey protein to total protein ratio as determined by capillary electrophoresis. *J. Dairy Sci.* **2003**, *86*, 2813–2817. [CrossRef]
31. Albillos, S.M.; Busto, M.D.; Perez-Mateos, M.; Ortega, N. Analysis by capillary electrophoresis of the proteolytic activity of a Bacillus subtilis neutral protease on bovine caseins. *Int. Dairy J.* **2007**, *17*, 1195–1200. [CrossRef]
32. Heck, J.M.L.; Olieman, C.; Schennink, A.; van Valenberg, H.J.F.; Visker, M.H.P.W.; Meuldijk, R.C.R.; van Hooijdonk, A.C.M. Estimation of variation in concentration, phosphorylation and genetic polymorphism of milk proteins using capillary zone electrophoresis. *Int. Dairy J.* **2008**, *18*, 548–555. [CrossRef]
33. Bezerra, T.K.A.; Araújo, A.R.R.; de Oliveira Arcanjo, N.M.; da Silva, F.L.H.; do Egypto Queiroga, R.d.C.R.; Madruga, M.S. Optimization of the HS-SPME-GC/MS technique for the analysis of volatile compounds in caprine Coalho cheese using response surface methodology. *Food Sci. Technol.* **2016**, *36*, 103–110. [CrossRef]
34. Lee, J.H.; Diono, R.; Kim, G.Y.; Min, D.B. Optimization of solid phase microextraction analysis for the headspace volatile compounds of Parmesan cheese. *J. Agric. Food Chem.* **2003**, *51*, 1136–1140. [CrossRef] [PubMed]
35. ISO (International Organisation for Standardisation). *Sensory Analysis - General Guidance for the Design of Test Rooms*; ISO 8589. 2007; ISO: Geneva, Switzerland, 2007.
36. Rehn, U.; Petersen, M.A.; Saedén, K.H.; Ardö, Y. Ripening of extra-hard cheese made with mesophilic DL-starter. *Int. Dairy J.* **2010**, *20*, 844–851. [CrossRef]
37. Walstra, P.; Noomen, A.; Geurts, T.J. Dutch-type varieties. In *Cheese: Chemistry, Physics and Microbiology. Major Cheese Groups*; Fox, P.F., Ed.; Chapman and Hall: London, UK, 1999; pp. 39–82.

38. Van den Berg, G.; Meijer, W.C.; Düsterhöft, E.M.; Smit, G. Gouda and related cheeses. In *Cheese: Chemistry, Physics and Microbiology*; Fox, P.F., McSweeney, P.L.H., Coagan, T.M., Guinee, T., Eds.; Elsevier: Oxford, UK, 2004; Volume 2, pp. 103–140.
39. Upadhyay, V.K.; McSweeney, P.L.H.; Magboul, A.A.A.; Fox, P.F. Proteolysis in cheese during ripening. *Cheese Chem. Phys. Microbiol.* **2004**, *1*, 37–41.
40. Hayes, M.G.; Oliveira, J.C.; McSweeney, P.L.H.; Kelly, A.L. Thermal inactivation of chymosin during cheese manufacture. *J. Dairy Res.* **2002**, *69*, 269–279. [CrossRef] [PubMed]
41. Sheehan, J.J.; Oliveira, J.C.; Kelly, A.L.; Mc Sweeney, P.L.H. Effect of cook temperature on primary proteolysis and predicted residual chymosin activity of a semi-hard cheese manufactured using thermophilic cultures. *Int. Dairy J.* **2007**, *17*, 826–834. [CrossRef]
42. Sousa, M.; Ardö, Y.; McSweeney, P.L.H. Advances in the study of proteolysis during cheese ripening. *Int. Dairy J.* **2001**, *11*, 327–345. [CrossRef]
43. Nielsen, S.S. Plasmin system and microbial proteases in milk: Characteristics, roles, and relationship. *J. Agric. Food Chem.* **2002**, *50*, 6628–6634. [CrossRef]
44. McSweeney, P.L.H.; Sousa, M.J. Biochemical pathways for the production of flavour compounds in cheeses during ripening: A review. *Lait* **2000**, *80*, 293–324. [CrossRef]
45. Fox, P.F.; Wallace, J.M. Formation of flavor compounds in cheese. *Adv. Appl. Microbiol.* **1997**, *45*, 17–85.
46. Ardö, Y.; Thage, B.V.; Madsen, J.S. Dynamics of free amino acid composition in cheese ripening. *Aust. J. Dairy Technol* **2002**, *57*, 109–115.
47. Ardö, Y. Flavour formation by amino acid catabolism. *Biotechnol. Adv.* **2006**, *24*, 238–242. [CrossRef] [PubMed]
48. Li, H.; Cao, Y. Lactic acid bacterial cell factories for gamma-aminobutyric acid. *Amino Acids* **2010**, *39*, 1107–1116. [CrossRef] [PubMed]
49. Yvon, M.; Rijnen, L. Cheese flavour formation by amino acid catabolism. *Int. Dairy J.* **2001**, *11*, 185–201. [CrossRef]
50. Jo, Y.; Benoist, D.M.; Ameerally, A.; Drake, M.A. Sensory and chemical properties of Gouda cheese. *J. Dairy Sci.* **2017**, 1–23. [CrossRef]
51. Van Leuven, I.; Van Caelenberg, T.; Dirinck, P. Aroma characterisation of Gouda-type cheeses. *Int. Dairy J.* **2008**, *18*, 790–800. [CrossRef]
52. Frank, D.C.; Owen, C.M.; Patterson, J. Solid phase microextraction (SPME) combined with gas-chromatography and olfactometry-mass spectrometry for characterization of cheese aroma compounds. *LWT-Food Sci. Technol.* **2004**, *37*, 139–154. [CrossRef]
53. Bellesia, F.; Pinetti, A.; Pagnoni, U.M.; Rinaldi, R.; Zucchi, C.; Caglioti, L.; Palyi, G. Volatile components of Grana Parmigiano-Reggiano type hard cheese. *Food Chem.* **2003**, *83*, 55–61. [CrossRef]
54. Tansman, G.F.; Kindstedt, P.S.; Hughes, J.M. Crystal fingerprinting: elucidating the crystals of Cheddar, Parmigiano-Reggiano, Gouda, and soft washed-rind cheeses using powder x-ray diffractometry. *Dairy Sci. Technol.* **2015**, *95*, 651–664. [CrossRef] [PubMed]

Article

On the Use of Ultrafiltration or Microfiltration Polymeric Spiral-Wound Membranes for Cheesemilk Standardization: Impact on Process Efficiency

Julien Chamberland [1,*], Dany Mercier-Bouchard [1], Iris Dussault-Chouinard [1], Scott Benoit [1], Alain Doyen [1], Michel Britten [2] and Yves Pouliot [1]

1 STELA Dairy Research Center, Institute of Nutrition and Functional Foods (INAF), Department of Food Sciences, Université Laval, Québec, QC G1V 0A6, Canada; dany.mercier-bouchard.1@ulaval.ca (D.M.-B.); iris.dussault-chouinard.1@ulaval.ca (I.D.-C.); scott.benoit.1@ulaval.ca (S.B.); alain.doyen@fsaa.ulaval.ca (A.D.); yves.pouliot@fsaa.ulaval.ca (Y.P.)

2 Food Research and Development Center (FRDC), Agriculture and Agri-Food Canada, Saint-Hyacinthe, QC J2S 8E3, Canada; michel.britten@canada.ca

* Correspondence: julien.chamberland.1@ulaval.ca; Tel.: +1-418-656-5988

Received: 8 May 2019; Accepted: 6 June 2019; Published: 8 June 2019

Abstract: Ultrafiltration (UF) and microfiltration (MF) are widely-used technologies to standardize the protein content of cheesemilk. Our previous work demonstrated that protein retention of a 0.1-µm MF spiral-wound membrane (SWM) was lower, but close to that of a 10 kDa UF one. Considering that the permeability of MF membranes is expected to be higher than that of UF ones, it was hypothesized that the former could improve the efficiency of the cheesemaking process. Consequently, the objectives of this work were to compare 0.1-µm MF and 10 kDa UF spiral-wound membranes in terms of (1) hydraulic and separation performance, (2) energy consumption and fouling behavior, (3) cheesemaking efficiency of retentates enriched with cream, and (4) economic performance in virtual cheesemaking plants. This study confirmed the benefits of using MF spiral-wound membranes to reduce the specific energy consumption of the filtration process (lower hydraulic resistance and higher membrane permeability) and to enhance the technological performance of the cheesemaking process (higher vat yield, and protein and fat recoveries). However, considering the higher serum protein retention of the UF membrane and the low price of electricity in Canada, the UF scenario remained more profitable. It only becomes more efficient to substitute the 10 kDa UF SWM by the 0.1-µm MF when energy costs are substantially higher.

Keywords: efficiency; microfiltration; ultrafiltration; cheesemilk standardization; process simulation

---

1. Introduction

Dairy products can be manufactured through many processing itineraries, but the most eco-efficient (i.e., having the lowest environmental impact and providing higher incomes) has to be selected to improve the competitiveness of the industry. In the cheese sector, the use of membrane filtration processes to concentrate the milk prior cheesemaking represents one way to improve its profitability and plant capacity [1,2]. Ultrafiltration (UF) is a widely used membrane technology. It allows to concentrate all the milk proteins, notably caseins (CN) and serum proteins (SP), whereas lactose and minerals are collected in the permeate [3,4]. It leads to improved coagulation properties [5–7] and increases cheese yield through a higher retention of milk components in the curd [3,6,8].

Microfiltration (MF) is also used for the protein standardization of cheesemilk [3]. As UF membranes, MF ones retain CN in the retentate, but generally have a low SP retention coefficient because of its higher mean pore size [9]. Due to the presence of SP in the MF permeate, it is considered as a "clean whey" since it is free of residues from cheesemaking such as caseinomacropeptite (CMP), cheese fines, colorants, lactic acid, or starter cultures found in traditional whey [10]. Furthermore, SP are found in the permeate in their native form, allowing the production of highly valuable whey protein isolates (WPI) that can be sold for their technological (foaming and gelling) and nutritional properties [11,12].

In a previous study, the performance of both 0.1- and 0.2-µm microfiltration (MF) spiral-wound membranes (SWM) was compared in the context of cheesemilk standardization [13]. Mercier-Bouchard et al. [13] confirmed the low SP retention property of a 0.2-µm MF SWM, but observed that SP are highly retained with a 0.1-µm polymeric MF SWM. A similar conclusion was reported in the literature, revealing the higher SP retention with MF SWM than with ceramic uniform transmembrane pressure (UTP), ceramic graded permeability (GP) membranes or hollow fiber ones [14,15]. In fact, SWM MF membranes, notably the 0.1-µm pore size, suffer from much more severe deposit formation so that the retained casein micelle layer exerts the retention effect.

Considering the largest pore size of MF membranes (higher permeability and hydraulic performance) and the fact that a more valuable permeate could be obtained (even if diluted), it was hypothesized that the cheesemilk standardization could be more efficient with the use of a 0.1-µm polymeric MF SWM, exhibiting a UF-like behavior, instead of a traditional 10 kDa UF SWM. The filtration performance of both membrane types was evaluated in the total recirculation and concentration modes at the pilot scale with skim milk at 50 °C. The cheesemaking efficiency of retentates collected during both filtration methods was also evaluated and compared with that of unconcentrated milk, and an economic assessment of the three scenarios tested was finally presented.

2. Materials and Methods

2.1. Raw Material

Pasteurized skim milk was purchased from a local dairy farm and stored at 4 °C until MF and UF experiments. These were performed in triplicate, in the recirculation mode, with the same batch of milk divided into three equal volumes of 300 L. For single-stage concentration and diafiltration (DF) modes, a different batch of milk was used for each membrane tested (0.1-µm MF and 10 kDa UF).

2.2. Filtration System

A pilot system described previously by Mercier-Bouchard et al. [13] was used for all the filtration experiments (model 393, Tetra Pak Filtration Systems, Champlin, MN, USA). Only one stage was used during this work with two MF or two UF membranes installed in series in the same loop. The 0.1-µm MF SWM was made of polyvinylidene fluoride (PVDF) (membrane v0.1, element specification model 3838, Synder Filtration, Vacaville, CA, USA), and the 10 kDa UF membrane was made of polyethersulfone (PES) (model DS-UH-3838, Microdyn-Nadir, Raleigh, NC, USA). They were mounted horizontally with respective surface areas of 13.38 m^2 and 10.14 m^2.

2.3. Operational Modes

2.3.1. Total Recirculation Mode

Filtration was performed at 50 °C in the total recirculation mode in order to determine the optimal transmembrane pressure (TMP) to be used in the concentration one. Experiments with 0.1-µm MF membranes were carried out at a TMP of 89.6, 106.9 and 124.1 kPa, as described in Mercier-Bouchard et al. [13], whereas values of 310.4, 379.4 and 447.5 kPa were applied on 10 kDa UF ones. Since neither type was operated with the same TMP, the permeability was chosen to compare them, as described by Methot-Hains et al. [16].

2.3.2. Concentration/DF of Skim Milk

Single-stage batch concentration and discontinuous DF with 1.5 diavolume (DV) were both performed at 50 °C at the optimal TMP, as determined in the recirculation mode, until reaching a targeted mass concentration factor (MCF) of 2.5×. Two DF were carried out (DF #1 and DF #2). Again, the permeability was used to compare the two membrane types. The true protein (TP) rejection coefficient was also calculated [17].

2.4. Membrane Fouling Characterization

The resistance-in-series model was applied to evaluate membrane fouling. The membrane resistance (R_m), reversible resistance (R_{rev}), irreversible resistance (R_{irrev}), and total resistance (R_{tot}) were calculated [4].

2.5. Chemical Analysis

Skim milk, retentate, permeate, and whey samples were analyzed according to the methodology described by Tremblay-Marchand et al. [17]. Briefly, the contents of the TP, CN, and non-protein nitrogen (NPN) were determined by the Kjeldahl digestion (AOAC International 991.20, 998.05, and 991.21, respectively). The total solids (TS) and fat (TF) were determined by the forced-air oven drying method (AOAC International 990.20) and the Mojonnier extraction one (AOAC International 989.05), respectively.

2.6. Energy Consumption

The electric energy consumption (Wh) of the MF and UF processes were obtained following the calculation of the power requirement (P, W) to operate the filtration system (Equation (1)).

$$P = \sqrt{3} \times U \times I \times \cos(\varphi) \quad (1)$$

where U, I and $\cos(\varphi)$ represent the voltage, current and power factor (0.65), respectively.

The current used to calculate P was that measured at the end of each concentration or diafiltration steps for the feed and recirculation pumps. The specific energy consumption (SEC, Wh per kilogram of permeate that is removed) was determined as follows (Equation (2)):

$$SEC = (P \times \Delta t)/V_P \quad (2)$$

where Δt and V_P represent the time needed to complete concentration and diafiltration steps, and the volume of permeate that is collected during them, respectively.

As in Mercier-Bouchard et al. [13], only electricity used to pump the fluids in the filtration system was considered.

2.7. Cheesemaking

The contents of TP of retentates during MF and UF were standardized to a final concentration of 7% (*w/w*) by reincorporating UF permeate. Fat standardization of retentates collected during both filtration types and of unconcentrated milk, was carried out by the addition of unpasteurized cream, obtained from a local dairy farm, in order to reach a final true protein to fat (TP/TF) ratio of 0.65. This low TP/TF corresponds to a high-fat cheesemilk [18]. Following standardization, the retentates were pasteurized at 68 °C for 30 min in a double-jacketed vessel mixer (model UMC-5, Stephan Machinery™, Hameln, Germany). The pH of the cheesemilks was adjusted with glucono-δ-lactone (GDL) to 6.50, and the model curds were produced according to the method described by Lauzin et al. [19] with the following modifications. At the end of the cheesemaking process, the curd was drained in two steps: in cotton cheesecloth over 30 min, and finally by centrifugation at 10,816× *g* for 30 min. The curds were vacuum-packaged and stored at 4 °C until chemical analysis. All experiments were performed in triplicate.

2.8. Cheesemaking Efficiency

The cheesemaking efficiency was evaluated with manufacturing yield (Y), moisture-adjusted yield (Y_{MA}), and protein and fat recovery (Y_P and Y_F, respectively). These variables were calculated based on the mass of the standardized cheesemilk (Vat) or one of the inputs (Inp). The Y (Equation (3)) and Y_{MA} (Equation (4)) were determined as follows, according to Guinee et al. [1]:

$$Y_{Vat}\ (\%) = 100 \times m_{curd}\ (kg)/m_{Vat}\ (kg) \tag{3}$$

where m_{curd} and m_{Vat} represent the mass of the curd and that of the standardized cheesemilk, respectively.

$$Y_{MAVat}\ (\%) = Y \times (100 - M_{curd})/(100 - M_{ref}) \tag{4}$$

where M_{curd} and M_{ref} represent the moisture (%) of the curd and that of a reference cheese, respectively. The latter corresponded to the cheese made from unconcentrated milk.

The protein recovery in the standardized cheesemilk (Y_{PVat}) (Equation (5)) was determined as follows:

$$Y_{PVat}\ (\%) = 100 \times (1 - (m_{Pw}/m_{PVat})) \tag{5}$$

where m_{Pw} and m_{PVat} represent the protein mass in the whey and that in the standardized cheesemilk.

The fat recovery in the standardized cheesemilk (Y_F) (Equation (6)) was determined as follows:

$$Y_{FVat}\ (\%) = 1 - (m_{Fw}/m_{FVat}) \tag{6}$$

where m_{FW} and m_{FVat} represent the fat mass in the whey and that in the standardized cheesemilk.

2.9. Economic Assessment

A process simulation was finally performed to compare the economic assessment of cheesemaking involving the three technological approaches presented previously: from standardized cheesemilk with MF or UF retentates, as well as from unconcentrated milk. It was done using the same parameters (i.e., membrane types, MCF, TP/TF ratio of the cheesemilk, number of DV or DF) as the experimental part, with the following modifications. Regarding the MF and UF scenarios, instead of concentrating the total volume of milk to a MCF of 2.5×, and to dilute it with the permeate, only a fraction was skimmed and concentrated. The retentate obtained was combined with cream and whole milk to standardize cheesemilk to a TP of 5.87% (*w/w*) and a TF of 9.03% (*w/w*). Missing data were interpolated from the ones obtained in the experimental part (i.e., membrane permeation fluxes at specific MCF). It was assumed that filtrations in virtual plants were performed in three-stage filtration systems.

The simulation considered a 1,000,000 kg daily delivery of whole raw milk (TP = 3.27% (*w/w*), TF = 3.97% (*w/w*), lactose = 4.81% (*w/w*), TS = 12.8% (*w/w*) [20]), cream (TP = 0.62% (*w/w*), and TF = 45.00% (*w/w*)) in virtual plants operating 265 days per year. Neither the cleaning or sanitation steps, nor the human resource requirements were taken into account. The simulation focused on the filtration and cheesemaking processes.

2.10. Statistical Analysis

Significant differences of each variable were detected by a one-way analysis of variance (one-way ANOVA) with RStudio software (v.1.1.463, [21]) using the Agricolae package (v.1.3-0, [22]). They were evaluated by the Fisher's least significant difference (LSD) test ($p < 0.05$).

3. Results

3.1. Effect of TMP on Normalized Permeation Flux in the Total Recirculation Mode

Three TMP were tested in the total recirculation mode for MF and UF of skim milk at 50 °C. As shown in Figure 1, no limiting flux was reached between 89.6 kPa and 124.1 kPa for the MF type, and between 310.4 kPa and 447.5 kPa for the UF one. Consequently, the highest TMP obtained for both membranes was selected for further concentration and diafiltration experiments. At these TMP, the permeation flux normalized per unit of pressure was of 0.34 and 0.13 kg h^{-1} m^{-2} kPa^{-1} for the MF and the UF membranes, respectively (Figure 1).

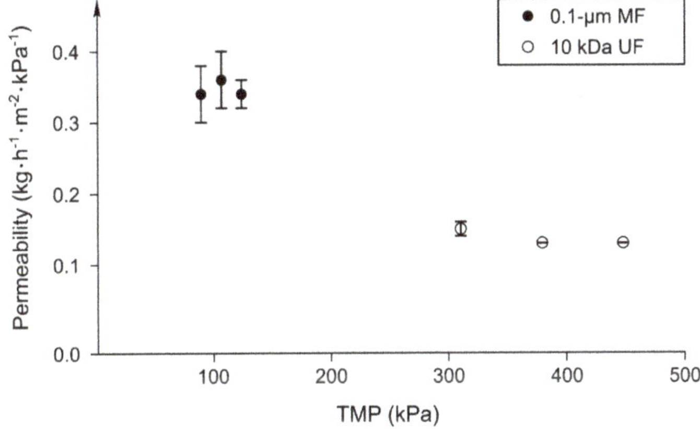

Figure 1. Permeability at different transmembrane pressures (TMP) during microfiltration (MF) and ultrafiltration (UF) of pasteurized skim milk (T = 50 °C) with 0.1-µm MF (●) and 10 kDa UF (○) membranes, respectively ($n = 3$, ± standard deviation [SD]).

3.2. Effect of Concentration and Diafiltration Modes on Normalized Permeation Flux

During concentration and DF of skim milk, the difference between the permeability of the membrane types varied significantly at all the MCF tested ($p < 0.05$) (Figure 2). The DF performed with both membranes increased the initial permeability of each reconcentration step. For example, the value for the MF type was of 0.39 kg h^{-1} m^{-2} kPa^{-1} and 0.44 kg h^{-1} m^{-2} kPa^{-1} at the beginning of the first and second DF, respectively, while the initial one was of 0.30 kg h^{-1} m^{-2} kPa^{-1} (Figure 2). The permeability decreased with the MCF during the concentration and both DF steps with the MF and UF membranes, but the flux reduction rate was similar in all the processing steps (Figure 2). It was, however, greater during MF with values between 34.4% (DF #2) and 38.2% (concentration), whereas those obtained during UF were between 13.4% (DF #2) and 31.5% (concentration) (Figure 2).

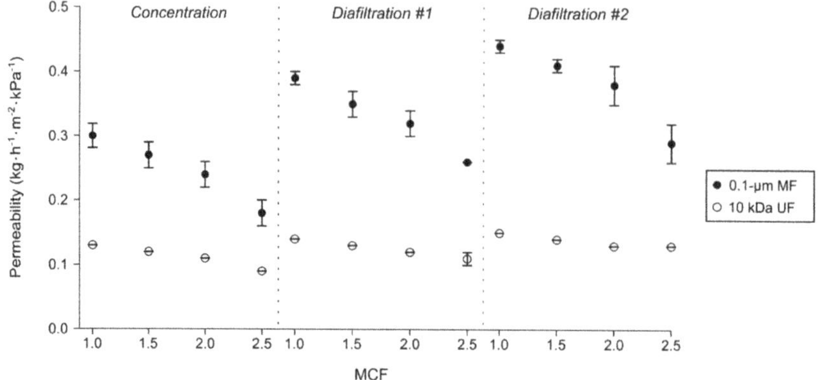

Figure 2. Normalized permeate flux as a function of the mass concentration factor (MCF) with 0.1-µm MF and 10 kDa UF membranes during concentration and two subsequent diafiltration steps (DF) of pasteurized skim milk at 50 °C ($n = 3$, ±SD).

3.3. Effect of Concentration and Diafiltration Modes on Retentate Composition

Globally, the UF membrane had a higher TP rejection coefficient ($p < 0.05$) (Table 1). Consequently, the TP content of permeates collected during MF was always significantly higher than the those during UF ($p < 0.05$), but no significant difference was found regarding the TP of 2.5× UF and MF retentates after the second DF ($p > 0.05$) (Table 1). The transmission of TP in the permeate during MF corresponded to around 11% of the TP of the skim milk. The DF effectively reduced the TS content of 2.5× retentates (MF and UF). For example, the TS content of the retentate during MF was reduced from 14.97 ± 0.65% (*w/w*) after the first concentration step to 11.48 ± 0.42% and 9.88 ± 0.02% (*w/w*) following the first and the second DF, respectively ($p < 0.05$) (Table 1). After the second DF, the 2.5× retentates (UF and MF) had the same TS content (10.09 ± 1.77% and 9.88 ± 0.02% (*w/w*), respectively) ($p > 0.05$) (Table 1). Even if the retentates were diafiltered, their contents of NPN did not differ significantly at the end of each concentration and DF steps ($p > 0.05$) (Table 1). The NPN concentrations were possibly too low to observe differences.

Table 1. Mean composition of skim milk retentates and permeates during MF and UF.

Sample	Membrane Type	Filtration Step	MCF	TP (% *w/w*)	NPN (%*w/w*)	TP Rejection Coefficient	Total Solids (%*w/w*)
Retentate	0.1-µm (MF)	Concentration	1.0×	3.13 ± 0.12 ^{cd}	0.15 ± 0.01 ^{a}	97.07 ± 0.75 ^{d}	8.77 ± 0.15 ^{de}
			2.5×	7.08 ± 1.62 ^{b}	0.09 ± 0.03 ^{abcd}		14.97 ± 0.65 ^{a}
		DF #1	1.0×	3.05 ± 0.22 ^{cd}	0.07 ± 0.01 ^{bcd}	97.11 ± 0.52 ^{d}	5.25 ± 0.17 ^{g}
			2.5×	8.94 ± 0.45 ^{a}	0.13 ± 0.06 ^{abc}		11.48 ± 0.42 ^{b}
		DF #2	1.0×	2.92 ± 0.11 ^{d}	0.05 ± 0.01 ^{d}	98.92 ± 0.21 ^{bc}	4.56 ± 0.87 ^{g}
			2.5×	8.35 ± 0.08 ^{ab}	0.14 ±0.11 ^{ab}		9.88 ± 0.02 ^{cd}
	10 kDa (UF)	Concentration	1.0×	3.53 ± 0.56 ^{cd}	0.13 ± 0.02 ^{abc}	99.36 ± 0.06 ^{ab}	8.48 ± 0.11 ^{ef}
			2.5×	8.91 ± 0.82 ^{a}	0.14 ± 0.02 ^{ab}		16.12 ± 0.05 ^{a}
		DF #1	1.0×	4.40 ± 1.22 ^{c}	0.07 ± 0.01 ^{bcd}	98.38 ± 0.13 ^{ab}	7.25 ± 1.01 ^{f}
			2.5×	9.23 ± 0.27 ^{a}	0.09 ± 0.03 ^{abcd}		11.96 ± 0.74 ^{b}
		DF #2	1.0×	3.67 ± 0.48 ^{cd}	0.06 ± 0.04 ^{cd}	99.82 ± 0.04 ^{a}	4.92 ± 0.57 ^{g}
			2.5×	9.15 ± 1.50 ^{a}	0.11 ± 0.07 ^{abcd}		10.09 ± 1.77 ^{c}
Permeate	0.1-µm (MF)	Concentration	2.5×	0.22 ± 0.11 ^{a}	0.09 ± 0.06 ^{ab}	-	5.93 ± 0.22 ^{a}
		DF #1	2.5×	0.26 ± 0.04 ^{a}	0.15 ± 0.08 ^{a}	-	2.06 ± 0.03 ^{b}
		DF #2	2.5×	0.09 ± 0.02 ^{b}	0.06 ± 0.02 ^{bc}	-	0.76 ± 0.01 ^{c}
	10 kDa (UF)	Concentration	2.5×	N.D.	0.15 ± 0.01 ^{a}	-	5.50 ± 0.06 ^{a}
		DF #1	2.5×	N.D.	0.05 ± 0.01 ^{bc}	-	1.41 ± 0.77 ^{bc}
		DF #2	2.5×	N.D.	0.02 ± 0.00 ^{c}	-	0.92 ± 0.56 ^{c}

$^{a-h}$ Values in the same column with a common superscript are not significantly different (LSD least significant difference test, $p > 0.05$, $n = 3$, ±SD). Retentates and permeates were compared separately. TP: true protein (TN−NPN); NPN: non-protein nitrogen × 6.38; TN: total nitrogen × 6.38; DF: diafiltration, N.D.: not detected.

3.4. Effect of Concentration and Diafiltration Modes on Energy Consumption

UF was more energy-demanding than MF, during the concentration or both diafiltration steps ($p < 0.05$) (Table 2). Considering the whole process, UF had a SEC of 13.31 ± 0.23 Wh per kilogram of permeate that is removed, whereas that of MF was of 8.59 ± 0.50 Wh per kilogram of permeate removed (Table 2). During MF, the concentration step was more energy-demanding than both diafiltration steps ($p < 0.05$), but the electricity used for these was similar ($p > 0.05$) (Table 2). During UF, the electricity consumption of each step was significantly different ($p > 0.05$) (Table 2).

Table 2. Energy consumption of feed and recirculation pumps during MF [1] and UF [1] of skim milk at 50 °C.

Filtration Step	Specific Energy Consumption [1] (Wh per kg of Permeate Removed)	
	MF	UF
Concentration	10.81 ± 0.77 ^{bA}	15.60 ± 0.28 ^{aA}
Diafiltration #1	8.05 ± 0.40 ^{bB}	12.89 ± 0.33 ^{aB}
Diafiltration #2	6.92 ± 0.51 ^{bB}	11.47 ± 0.10 ^{aC}
Whole process	8.59 ± 0.50 ^{b}	13.31 ± 0.23 ^{a}

[1] The optimal transmembrane pressure (TMP) was used to perform MF (124.1 kPa) and UF (448.2 kPa) until reaching a mass concentration factor (MCF) of 2.5× during the concentration and diafiltration steps ($n = 3$, ± SD). ^{a,b} Values in the same row with a tiny common superscript letter are not significantly different (one-way ANOVA, $p > 0.05$, $n = 3$). $^{A–C}$ Values in the same column with a common capital superscript letter are not significantly different (LSD least significant difference test, $p > 0.05$, $n = 3$).

3.5. Effect of Concentration and Diafiltration Modes on Membrane Fouling

The MF and UF membranes had a similar R_{rev} ($p > 0.05$) (Table 3). However, the latter had a higher R_m, R_{irrev} and R_{tot} than the former ($p < 0.05$) (Table 3). The R_{tot} of the UF membrane type was 2.25 times higher than that of the MF one (Table 3). The R_{irrev} of the UF membranes represented 42% of its R_{tot}, whereas that of the MF membranes represented 26% of its R_{tot}. Furthermore, the fouling type of the latter was rather reversible than irreversible. Its R_{rev} was 2.31 times higher than its R_{irrev}, while it was only 1.13 times higher with the UF membrane (Table 3).

Table 3. Hydraulic resistance of MF and UF membranes following concentration and diafiltration processes.

Membrane Type	Resistance Type (10^{-13} m^{-1})			
	R_m	R_{rev}	R_{irrev}	R_{tot}
MF (0.1-µm)	0.47 ± 0.01 ^{b}	1.87 ± 0.92 ^{a}	0.81 ± 0.13 ^{b}	3.15 ± 0.96 ^{b}
UF (10 kDa)	0.72 ± 0.05 ^{a}	3.38 ± 0.30 ^{a}	3.00 ± 0.39 ^{a}	7.09 ± 0.23 ^{a}

$^{a–b}$ Means in the same column with a common superscript are not significantly different ($p > 0.05$). R_m: membrane resistance, R_{rev}: reversible resistance, R_{irrev}: irreversible resistance, R_{tot}: total resistance.

3.6. Cheesemaking Efficiency of Unconcentrated Cheesemilk, and Standardized with MF and UF Retentates

The higher moisture-adjusted yield (Y_{MAVat}) was obtained with the MF retentate (29.67 ± 0.51%, $p < 0.05$), but the use of UF also permitted to increase Y_{MAVat} compared to unconcentrated cheesemilk ($p < 0.05$) (Table 4). The MF retentate also allowed higher cheesemilk protein (Y_{PVat}) and fat (Y_{FVat}) recoveries in cheese (90.73 ± 0.13% and 96.91 ± 0.26%, respectively), which also significantly increased the FDM ratio ($p < 0.05$) (Table 4). The use of the same TP/TF ratio in the cheesemilk in the three scenarios allowed a similar draining behavior, revealed by a similar MNFS ratio (67.69 to 67.85) between the three cheeses ($p > 0.05$) (Table 4).

Table 4. Cheesemaking efficiency of MF and UF retentates obtained with spiral-wound membranes.

Indicator	Unconcentrated Cheesemilk	0.1-µm MF Retentate	10 kDa UF Retentate
TS (%)	51.11 ± 0.23 [b]	53.30 ± 1.29 [a]	52.09 ± 0.50 [ab]
MNFS (%)	67.69 ± 0.61 [a]	67.85 ± 1.53 [a]	67.75 ± 0.67 [a]
FDM (%)	54.34 ± 0.92 [c]	58.50 ± 0.93 [a]	56.23 ± 0.87 [b]
Y_{Vat} (%)	14.13 ± 0.25 [b]	28.46 ± 0.53 [a]	28.21 ± 0.47 [a]
Y_{MAVat} (%)	14.13 ± 0.30 [c]	29.67 ± 0.51 [a]	28.75 ± 0.42 [b]
Y_{FVat} (%)	89.46 ± 0.41 [c]	96.91 ± 0.26 [a]	96.29 ± 0.12 [b]
Y_{PVat} (%)	76.15 ± 0.26 [c]	90.73 ± 0.13 [a]	86.17 ± 0.20 [b]

[a–c] Values in the same rows with a common superscript are not significantly different (Fisher's LSD test, $n = 3$, ±SD, $p > 0.05$). TS: total solids, MNFS: moisture in nonfat substances, FDM: fat content in dry matter, Y_{Vat}: mass yield reported on the mass of cheesemilk in vats, Y_{MA}: moisture-adjusted yield with the moisture content of the cheese made from unconcentrated milk, Y_{FV}: fat recovery from the standardized cheesemilk, Y_{PV}: protein recovery from the standardized cheesemilk.

3.7. Economic Assessment of MF and UF Approaches through a Process Simulation

The results obtained in the experimental part were used to perform a process simulation in virtual plants receiving 1,000,000 kg of whole raw milk daily (Tables 5 and 6). Similar patterns between the Y_{MA} observed in the experimental part and the predicted Y_{Vat} (calculated with cheeses having the same moisture content) presented in Table 5 were observed. It is, however, important to mention that the Y_{Vat} predicted in the process simulation was lower (i.e., by 1.16% for the UF approach); the difference between Y_{Vat} of MF and UF scenarios was different as well.

Table 5. Mass balance obtained from a process simulation of cheesemaking approaches involving milk concentration or not in virtual plants receiving 1,000,000 kg of whole milk daily.

Indicator	Scenarios		
	Unconcentrated Milk	MF-Standardized Cheesemilk	UF-Standardized Cheesemilk
Inputs (kg)			
Whole milk	1,000,000	1,000,000	1,000,000
Cream	24,070	23,178	30,012
Calcium chloride	143	73	78
Starter culture [1]	20	10	11
Coagulant enzyme	232	119	127
Outputs (kg)			
Cheese [2]	145,049	147,531	154,592
Cheese whey	879,995	376,863	405,647
Permeate	-	602,466	628,049
Diafiltrate	-	1,204,932	1,256,099
Predicted mass yield [3] (%)			
Y_{Vat}	14.16	28.13	27.59
Y_{Inp}	14.16	14.42	15.01
Y_{PInp}	76.15	84.57	85.73

[1] Lyophilized culture. [2] Cheese had a total solid (TS) content of 51.11%, as cheese made with the unconcentrated milk in the experimental part. Resulting masses were obtained from the protein and fat recoveries measured in the experimental part. Lactose content had the same concentration in the aqueous phase of the cheese as in the cheesemilk. [3] The mass yields were identical to the moisture-adjusted ones since cheeses made by the three scenarios had a similar composition. Y_{Vat}: mass yield reported on the mass of cheesemilk in vats, Y_{Inp}: mass yield reported on the mass of inputs, namely raw milk and cream, Y_{PInp}: protein recovery from the inputs.

Table 6. Economic assessment calculated from a process simulation of cheesemaking approaches involving milk concentration or not in virtual plants receiving 1,000,000 kg of whole milk daily.

Indicator	Price	Scenarios		
		Unconcentrated Milk	**MF-Standardized Cheesemilk**	**UF-Standardized Cheesemilk**
Expenditures				
Inputs				
Whole milk	84.56 Can$ 100 kg^{-1}	845,600	845,600	845,600
Cream	3.95 Can$ kg^{-1}	95,077	91,553	118,549
Calcium chloride	2.23 Can$ kg^{-1}	320	164	175
Starter culture [1]	600 Can$ kg^{-1}	12,289	6293	6723
Coagulant enzyme	28.32 Can$ kg^{-1}	6583	3371	3602
Subtotal	Can$	959,869	946,981	974,649
Filtration				
Membrane replacement [2]	0.23 Can$ m^{-2} day^{-1}	0	1,169	714
Electricity	0.0327 Can$ kWh^{-1}	0	508	821
Power	0.42 Can$ kW^{-1}	0	821	1327
Subtotal	Can$	0	2498	2862
Operating costs	Can$	959,869	949,479	977,511
	Can$ kg of $cheese^{-1}$	6.62	6.44	6.32

[1] Lyophilized culture. [2] Membrane replacement cost considers a replacement once per year. Filtration processes were operated 8 h per day, 265 days per year, in a three-stage system.

The process simulation confirmed the interest of using filtration processes with the TP/TF ratio of the cheesemilk used in this study, to increase the manufacturing yields calculated from the mass of the inputs (Table 5), and to reduce the operating costs (Table 6). As a fraction of SP are transmitted in the MF permeate, a lower protein recovery from the inputs (Y_{PInp}) was observed in the MF scenario compared to the UF one (Table 5). It remained, however, higher than with the unconcentrated milk.

4. Discussion

Even though previous works confirmed the high SP retention coefficient value of MF SWM [13,23], notably in comparison with ceramic ones [14], that of the 10 kDa UF SWM was significantly higher in this study (Table 1). The MF membrane had the advantage of a lower hydraulic resistance (Table 3) and a higher permeability at every MCF tested (Figure 2), which was in accordance with its SEC that was low (Wh per kilogram of permeate collected) compared to the UF one (Table 2). As seen in the process simulation, the cheesemilk standardization through MF would require a larger membrane surface area, due to a lower permeation flux at its optimal TMP (despite a higher permeability), but would be less energy-demanding (Table 6).

In the cheesemaking facility, the fat-enriched MF retentate exhibited a higher efficiency in terms of vat yields, as well as protein and fat recoveries (Table 4). Indeed, UF- and MF-standardized milk had the same TP/TF ratio, but the latter retentate had a slightly higher CN/SP one, which possibly contributed to enhancing milk coagulation properties and increased its constituent recoveries [24–26]. Conversely, the standardized cheesemilk made from unconcentrated milk had the poorer performance. The same TP/TF should have led to a similar fat loss to other scenarios, but its lower TP content possibly led to a weaker structure of the protein matrix, which is thus conductive to a lower retention of fat (Y_{FVat}) [1]. The fat-enriched UF retentate showed a lower protein retention from the standardized cheesemilk (Y_{PVat}) compared to the MF one (Table 4), due to the higher content of SP lost in the whey. However, the UF-standardized milk exhibited a higher protein retention than the unconcentrated milk, because a lower volume of whey was generated, and with that a lower loss of SP, as confirmed previously for the manufacture of Cheddar cheese [1,27]. In other studies [1,28,29], the enrichment in proteins with MF or UF retentates did not increase their recovery from input material (milk, cream and dairy ingredients) to cheeses. However, the cheeses that were made during the present studies had a higher moisture content (lower volume of whey expelled) and a lower TP/TF ratio that was high, which possibly affected both protein and fat recoveries.

Overall, both scenarios involving cheesemilk standardization with dairy retentates had a lower operating cost per kilogram of cheese and would increase the margin of the cheesemaking process in comparison with the unconcentrated scenario (Table 6), as also predicted by Papadatos et al. [30]. However, three elements may have biased results: (1) the cheesemilks had a high fat content, which generated important cream expenditures; (2) they also had a different CN content, due to the use of a TP/TF ratio; and (3) the cheesemaking performance of the MF scenario was evaluated directly on the retentate instead of the milk standardized with the latter, as generally performed in the industry [28]. The first two elements appeared more important, as the UF scenario remained the more efficient even with a possible overestimation of the MF cheesemilk performance. Indeed, the 2.5× UF retentate had a higher TP content than the MF one due to the transmission of SP in the MF permeate. Consequently, a higher volume of cream input (6485 kg) was needed in the UF scenario. It increased the operating cost of the UF process by 26,996 Can$ (Table 6). In fact, no bias would occur with the use of a CN/TF ratio, since MF and UF have a similar CN content. This statement was confirmed by an uncertainty analysis performed with a CN/TF instead of a TP/TF ratio of 0.65 (Table 7). With the CN-based standardization target, only slight differences were observed between MF and UF scenarios in terms of inputs or filtration expenditures. The UF scenario was however the more efficient with an operating cost (per kilogram of cheese) 0.5% lower than the MF one, and with a higher Y_{Inp} (Table 7), again due to its higher SP retention. In countries where the energy price is higher, the conclusion could be different. In fact, with the Canadian milk price, the MF scenario would become the more profitable if the one for electricity was 6.4 times higher.

Table 7. Uncertainty analysis using a CN/TF ratio of 0.65 instead of a TP/TF ratio.

Indicator	Price	Scenarios		
		Unconcentrated Milk	MF-Standardized Cheesemilk	UF-Standardized Cheesemilk
Expenditures				
Inputs	Can$	879,859	891,849	895,681
Filtration	Can$	0	2431	2815
Operating costs	Can$	879,859	894,280	898,496
	Can$ kg of cheese^{-1}	6.87	6.55	6.52
Y_{Inp}	%	12.85	13.53	13.65
Y_{PInp}	%	76.15	85.10	86.17

Y_{Inp}: mass yield reported on the mass of inputs, Y_{PInp}: protein recovery from the inputs.

5. Conclusions

Contrarily to the previous hypothesis proposed by Mercier-Bouchard et al. [13], even if the MF scenario was the less energy-demanding, notably because of the higher permeability of MF membranes, and had better cheesemaking performance of the MF retentate (calculated from the cheesemilk data collected), the UF one remained more efficient. In fact, with the low cost of electricity in the Canadian economic context, the gain associated with the lower use of energy with the 0.1-μm MF SWM was not sufficient to compensate the loss of SP in the MF permeate, as low as it was. By focusing only on the milk standardization and cheesemaking steps, the 0.1-μm MF SWM could not substitute the 10 kDa UF one without a substantial increase in energy costs (i.e., with energy costs at least 6.4 times higher). A further study considering the processing costs of cheese whey, MF and UF permeates as well as diafiltrates will be necessary to draw definitive conclusions on the economic assessments.

It should help to determine if processing diluted MF permeates generated from 0.1-μm MF SWM could contribute to increase the profitability of the MF process over the UF one.

Author Contributions: J.C. wrote the manuscript and carried out the statistical analysis. D.M.-B. designed and performed the filtration experiments. I.D.-C. manufactured the model cheeses. J.C. did the economic assessments. All authors discussed the results and commented on the manuscript at all stages. Y.P. acquired the funding.

Funding: This research was financed by the NSERC-Novalait Industrial Research Chair on Process Efficiency in Dairy Technology, grant number IRCPJ 46130-12 to Yves Pouliot.

Acknowledgments: The authors are thankful to Dominique Fournier for editing this manuscript as well as to Diane Gagnon, Mélanie Martineau, and Pascal Lavoie (Department of Food Sciences, Université Laval) for their technical assistance during the experiments and chemical analysis.

Conflicts of Interest: The authors declare no conflict of interest.

References

1. Guinee, T.P.; O'Kennedy, B.T.; Kelly, P.M. Effect of milk protein standardization using different methods on the composition and yields of Cheddar cheese. *J. Dairy Sci.* **2006**, *89*, 468–482. [CrossRef]
2. Kosikowski, F. New cheese-making procedures utilizing ultrafiltration. *Food Technol.* **1986**, *40*, 71–77.
3. Mistry, V.V.; Maubois, J.-L. Application of membrane separation technology to cheese production. In *Cheese: Chemistry, Physics and Microbiology*, 4th ed.; McSweeney, P.L.H., Fox, P.F., Cotter, P.D., Everett, D.W., Eds.; Academic Press: Cambridge, MA, USA, 2017; pp. 677–697.
4. Cheryan, M. *Ultrafiltration and Microfiltration Handbook*, 2nd ed.; CRC Press: Boca Raton, FL, USA, 1998.
5. Guinee, T.P.; Gorry, C.B.; Callaghan, D.J.O.; Brendan, T.; Kennedy, O.; Brien, N.O.; Fenelon, M.A. The effects of composition and some processing treatments on the rennet coagulation properties of milk. *Int. J. Dairy Technol.* **1997**, *50*, 99–106. [CrossRef]
6. Govindasamy-Lucey, S.; Jaeggi, J.J.; Martinelli, C.; Johnson, M.E.; Lucey, J.A. Standardization of milk using cold ultrafiltration retentates for the manufacture of Swiss cheese: Effect of altering coagulation conditions on yield and cheese quality. *J. Dairy Sci.* **2011**, *94*, 2719–2730. [CrossRef] [PubMed]
7. Garnot, P. Influence of milk concentration by UF on enzymatic coagulation. *Bull. Int. Dairy Fed.* **1988**, *225*, 11–15.
8. Hu, K.; Dickson, J.M.; Kentish, S.E. Microfiltration for casein and serum protein separation. In *Membrane Processing for Dairy Ingredient Separation*; Hu, K., Dickson, J.M., Eds.; John Wiley & Sons: Hoboken, NZ, USA, 2015; pp. 1–34.
9. Fauquant, J.; Maubois, J.L.; Pierre, A. Microfiltration du lait sur membrane minérale. *Tech. Laitière Mark.* **1988**, *1028*, 21–23.
10. Ardisson-Korat, A.V.; Rizvi, S.S.H. Vatless manufacturing of low-moisture part-skim mozzarella cheese from highly concentrated skim milk microfiltration retentates. *J. Dairy Sci.* **2004**, *87*, 3601–3613. [CrossRef]
11. Maubois, J.L. Membrane microfiltration: A tool for a new approach in dairy technology. *Aust. J. Dairy Technol.* **2002**, *57*, 92–96.
12. Gésan-Guiziou, G. Integrated membrane operations in whey processing. In *Integrated Membrane Operations in the Food Production*; Cassano, A., Drioli, E., Eds.; De Gruyter: Berlin, Germany, 2014; pp. 133–146.
13. Mercier-Bouchard, D.; Benoit, S.; Doyen, A.; Britten, M.; Pouliot, Y. Process efficiency of casein separation from milk using polymeric spiral-wound microfiltration membranes. *J. Dairy Sci.* **2017**, *100*, 8838–8848. [CrossRef]
14. Zulewska, J.; Newbold, M.; Barbano, D.M. Efficiency of serum protein removal from skim milk with ceramic and polymeric membranes at 50 °C. *J. Dairy Sci.* **2009**, *92*, 1361–1377. [CrossRef]
15. Heino, A. Microfiltration in Cheese and Whey Processing. Master's Thesis, University of Helsinki, Helsinki, Finland, 2010.
16. Méthot-Hains, S.; Benoit, S.; Bouchard, C.; Doyen, A.; Bazinet, L.; Pouliot, Y. Effect of transmembrane pressure control on energy efficiency during skim milk concentration by ultrafiltration at 10 and 50 °C. *J. Dairy Sci.* **2016**, *99*, 8655–8664. [CrossRef] [PubMed]
17. Tremblay-Marchand, D.; Doyen, A.; Britten, M.; Pouliot, Y. A process efficiency assessment of serum protein removal from milk using ceramic graded permeability microfiltration membrane. *J. Dairy Sci.* **2016**, 1–14. [CrossRef] [PubMed]

18. Guinee, T.P.; Mulholland, E.O.; Kelly, J.; Callaghan, D.J.O. Effect of protein-to-fat ratio of milk on the composition, manufacturing efficiency, and yield of Cheddar cheese. *J. Dairy Sci.* **2007**, *90*, 110–123. [CrossRef]
19. Lauzin, A.; Dussault-Chouinard, I.; Britten, M.; Pouliot, Y. Impact of membrane selectivity on the compositional characteristics and model cheese-making properties of liquid pre-cheese concentrates. *Int. Dairy J.* **2018**, *83*, 34–42. [CrossRef]
20. Fox, P.F.; McSweeney, P.L.H.; Cogan, T.M.T.; Guinee, T.P. *Cheese—Chemistry, Physics and Microbiology*; Academic Press: Cambridge, MA, USA, 2004.
21. Racine, J.S. RStudio: A platform-independent IDE for R and Sweave. *J. Appl. Econom.* **2012**, *27*, 167–172. [CrossRef]
22. De Mendiburu, F. Agricolae: Statistical procedures for agricultural research. *R Packag. Vers.* **2014**, *1*, 1–6.
23. Beckman, S.L.; Barbano, D.M. Effect of microfiltration concentration factor on serum protein removal from skim milk using spiral-wound polymeric membranes. *J. Dairy Sci.* **2013**, *96*, 6199–6212. [CrossRef] [PubMed]
24. Daviau, C.; Famelart, M.-H.; Pierre, A.; Goudédranche, H.; Maubois, J.-L. Rennet coagulation of skim milk and curd drainage: Effect of pH, casein concentration, ionic strength and heat treatment. *Lait* **2000**, *80*, 397–415. [CrossRef]
25. Pierre, A.; Fauquant, J.; Le Graet, Y.; Piot, M.; Maubois, J.L. Préparation de phosphocaséinate natif par microfiltration sur membrane. *Lait* **1992**, *72*, 461–474. [CrossRef]
26. Caron, A.; St-Gelais, D.; Pouliot, Y. Coagulation of milk enriched with ultrafiltered or diafiltered microfiltered milk retentate powders. *Int. Dairy J.* **1997**, *7*, 445–451. [CrossRef]
27. Oommen, B.S.; Mistry, V.V.; Nair, M.G. Effect of homogenization of cream on composition, yield, and functionality of Cheddar cheese made from milk supplemented with ultrafiltered milk. *Lait* **2000**, *80*, 77–91. [CrossRef]
28. Heino, A.; Uusi-Rauva, J.; Outinen, M. Pre-treatment methods of Edam cheese milk. Effect on cheese yield and quality. *LWT-Food Sci. Technol.* **2010**, *43*, 640–646. [CrossRef]
29. Neocleous, M.; Barbano, D.M.; Rudan, M.A. Impact of low concentration factor microfiltration on milk component recovery and Cheddar cheese yield. *J. Dairy Sci.* **2002**, *85*, 2415–2424. [CrossRef]
30. Papadatos, A.; Neocleous, M.; Berger, A.M.; Barbano, D.M. Economic feasibility evaluation of microfiltration of milk prior to cheesemaking. *J. Dairy Sci.* **2003**, *86*, 1564–1577. [CrossRef]

Article

Exploring the Use of a Modified High-Temperature, Short-Time Continuous Heat Exchanger with Extended Holding Time (HTST-EHT) for Thermal Inactivation of Trypsin Following Selective Enzymatic Hydrolysis of the β-Lactoglobulin Fraction in Whey Protein Isolate

Laura Sáez [1,2,3], Eoin Murphy [1,3], Richard J. FitzGerald [2] and Phil Kelly [1,*]

1 Teagasc Food Research Centre, Moorepark, Fermoy, P61 C996 Co. Cork, Ireland
2 Department of Biological Sciences, University of Limerick, V94 T9PX Limerick, Ireland
3 Food for Health Ireland, Teagasc Food Research Centre, Moorepark, Fermoy, P61 C996 Co. Cork, Ireland
* Correspondence: philk51@hotmail.com; Tel.: +353-877735352

Received: 28 July 2019; Accepted: 21 August 2019; Published: 26 August 2019

Abstract: Tryptic hydrolysis of whey protein isolate under specific incubation conditions including a relatively high enzyme:substrate (E:S) ratio of 1:10 is known to preferentially hydrolyse β-lactoglobulin (β-LG), while retaining the other major whey protein fraction, i.e., α-lactalbumin (α-LA) mainly intact. An objective of the present work was to explore the effects of reducing E:S (1:10, 1:30, 1:50, 1:100) on the selective hydrolysis of β-LG by trypsin at pH 8.5 and 25 °C in a 5% (*w*/*v*) WPI solution during incubation periods ranging from 1 to 7 h. In addition, the use of a pilot-scale continuous high-temperature, short-time (HTST) heat exchanger with an extended holding time (EHT) of 5 min as a means of inactivating trypsin to terminate hydrolysis was compared with laboratory-based acidification to <pH 3 by the addition of HCl, and batch sample heating in a water bath at 85 °C. An E:S of 1:10 resulted in 100% and 30% of β-LG and α-LA hydrolysis, respectively, after 3 h, while an E:S reduction to 1:30 and 1:50 led >90% β-LG hydrolysis after respective incubation periods of 4 and 6 h, with <5% hydrolysis of α-LA in the case of 1:50. Continuous HTST-EHT treatment was shown to be an effective inactivation process allowing for the maintenance of substrate selectivity. However, HTST-EHT heating resulted in protein aggregation, which negatively impacts the downstream recovery of intact α-LA. An optimum E:S was determined to be 1:50, with an incubation time ranging from 3 h to 7 h leading to 90% β-LG hydrolysis and minimal degradation of α-LA. Alternative batch heating by means of a water bath to inactivate trypsin caused considerable digestion of α-LA, while acidification to <pH 3.0 restricted subsequent functional applications of the protein.

Keywords: Hydrolysis; WPI; trypsin; α-lactalbumin; β-lactoglobulin; thermal inactivation

---

1. Introduction

Protein hydrolysis is a well-established process whereby proteins are cleaved to form peptides of different sizes under aqueous conditions. The nature of the peptides obtained depends upon the site(s) of protein cleavage, which in turn defines their length. Hydrolysis may be undertaken using alkali, acid or enzymes. An advantage of enzymatic hydrolysis is that the process usually takes place under relatively mild operating conditions—in contrast to alkali or acid hydrolysis, where extreme pH and temperature values may affect protein structure [1,2].

Enzymatic hydrolysis is commonly applied in many fields, i.e., waste treatment [3–5], detergent formulation [6] and at various stages during food processing. The products and ingredients derived from the hydrolysis of proteins in the food industry are of particular significance due to their biofunctional properties [7]. This is in addition to the enhanced technofunctional properties obtainable following enzymatic hydrolysis, e.g., enhanced solubility, foaming capacity, emulsion and gelation [8–13]. Other potential applications of enzymatic hydrolysis include the reutilisation of byproduct streams of food processing previously considered waste and a problem in terms of negative environmental impact [14,15]. A further field of interest for the food industry is the characterisation of the nutritional and bioactivity properties of peptides generated by enzymatic hydrolysis, which in recent years has highlighted potential biomarkers for physiological benefit such as antioxidant activity, regulation of gastric transit, antimicrobial activity, anticaries activity, antihypertensive activity, anti-inflammatory, satiety control and reduction of allergenic potential [7,16–23]. Reducing allergenic potential by enzymatic hydrolysis is especially important in the case of the dairy industry. Cow's milk allergy (CMA) is considered the most common allergy in children under three years of age. The reaction is triggered by one or more of the milk proteins, the most allergenic of which are represented in the following order: caseins, β-LG, α-LA [24]. Due to the heat-labile nature of whey proteins, their allergenicity may be modulated to some extent by certain thermal processing treatments, thus making their use in food formulations, in the form of whey protein concentrate (WPC) or whey protein isolate (WPI), preferential compared to the more heat-stable caseins. It is notable, however, that the β-LG fraction, which represents approximately 53% of the whey protein in cow's milk, is not present in human milk.

Enzyme activity ideally requires a combination of "optimum" conditions of temperature, pH, concentration of protein substrate and enzyme type [25–28] in order to maximise the degree of hydrolysis (DH or %DH). Thus, the optimum point of hydrolytic activity represents the set of conditions under which the highest conversion is achieved in a given time. Much work has focussed on maximising %DH in order to produce more and shorter peptides; however, there is increasing appreciation that intermediate %DH levels may have a more pronounced influence on final hydrolysate properties [29,30]. Furthermore, numerous substrate pre- or post-treatments have also been investigated in order to expand potential hydrolysate functionality [31], e.g., thermal pre-treatments [32], high pressure treatments [33] or a combination of endo- and exo-proteases [34].

Recent studies have shown that targeted or selective hydrolysis of proteins generates hydrolysates with specific peptides that have different properties to those of a fully hydrolysed protein [35–40]. The operating protocol necessary to achieve selective protein hydrolysis demands a specific set of incubation conditions, which may be less than optimum for the hydrolysis reaction per se [40]. In the case of WPI, a selective hydrolysis of β-LG may reduce the risk of allergy and at the same time enable the generation of a co-product enriched in α-LA—the predominant whey protein fraction in human milk, which is a key nutritional source of essential amino acids in sufficient concentration to promote infant development. Supplementation of infant formula with α-LA helps lower the total protein content of formula to resemble that of human milk. In addition, α-LA possesses a wide range of biological activities such as absorption of minerals, antibacterial activity, antioxidant activity, immunomodulatory effects and antitumor activity, which promote the health of the neonate [41,42]. The specific conditions reported for the selective hydrolysis of β-LG by trypsin or chymotrypsin include temperature (25 °C) and pH (pH 8.5), which resulted in minimal hydrolysis of the α-LA present [43–45].

At an industrial scale, thermal processes are frequently used to inactivate enzymes at the end of a reaction or on attaining a desired %DH. Lisak et al. [45] used non-thermal methods of inactivating chymotrypsin following WPI-based selective digestion of β-LG in order to avoid the risk of enzyme reactivation during time lags in thermal transfer buildup to the target temperature, resulting in excessive digestion of the major WPI fractions. Hence, the addition of inhibitors and acidification to <pH 3 were employed as alternative methods of enzyme inactivation. There are practical constraints, however, when using such approaches, i.e., cost, in the first instance; while heavily acidified substrates in the latter require neutralisation before further processing, thereby also potentially leading to enzyme reactivation.

In this study, a continuous high-temperature, short-time heater (HTST) with extended holding time (EHT)-based thermal treatment was assessed as a means of inactivating trypsin-based selective hydrolysis of β-LG in a WPI substrate dispersion while quantifying the formation of undesirable side reactions associated with protein denaturation and protein-peptide aggregation. The ultimate objective was to minimise possible aggregate formation that could interfere with subsequent research efforts aimed at producing an α-LA-enriched product from the targeted WPI hydrolysis. Furthermore, since the original research [43] using trypsin was conducted at a relatively high E:S of 1:10, a further objective was to evaluate a reduced E:S (down to 1:100) in terms of the incubation conditions necessary to maximise β-LG digestion while minimising the digestion of α-LA.

2. Material and Methods

2.1. Targeted Enzymatic Hydrolysis

Commercially sourced whey protein isolate (WPI) Isolac®, provided by Carbery Food Ingredients, Ballineen, Ireland, with a total protein content of 91.4% was dispersed at 5% (*w*/*v*) in distilled water using an electronic overhead stirrer model VWR VOS 40 digital (VWR International Ltd., Dublin, Ireland). The three-bladed propeller (55 mm in diameter) stirrer was set at 300 rpm for 30 min at room temperature during WPI reconstitution and the dispersions were allowed to stand overnight at 4 °C to allow complete hydration.

Enzymatic proteolysis was carried out using a commercial trypsin preparation (Trypsin 250, a porcine pancreas-based extract supplied by Biocatalysts, Cardiff, UK) with a specific activity of 250 U/mg protein. Incubation conditions were set at 25 °C and pH 8.5 in accordance with the conditions described [43] in order to pursue targeted hydrolysis of β-LG in the WPI dispersion. In order to further optimise this targeted enzymatic hydrolysis, samples were prepared at different enzyme:substrate ratios (E:S) (1:10, 1:30, 1:50, 1:100) (*w*:*w*) and over hydrolysis durations ranging from 1 to 7 h. Hydrolysate samples (5 mL) were taken hourly and the reaction was stopped by the addition of 1 mL of 1 M HCl in order to reduce to pH < 3 according to the enzyme supplier's protocol. The samples were frozen at −20 °C and retained for further analysis.

The enzymatic reaction was controlled using a Metrohm 842 Titrando pH-STAT titrator (Herisau, Switzerland), which maintained pH 8.5 by dosing 1M NaOH throughout the hydrolysis period.

2.2. Degree of Hydrolysis (%DH)

Quantification of %DH by the pH-stat method was carried out using the following formula:

$$DH\% = 100BN_b(1/\alpha)(1/MP)(1/h_{tot}),$$

where B is the mL of NaOH solution needed to maintain constant pH, N_b the normality of the alkaline solution, α the average degree of dissociation of the α-NH_2 groups (the value for α at pH 8.5 and 25 °C is 1.16), MP is the mass (g) of protein and h_{tot} the total number of peptide bonds in the protein substrate—the value used for whey protein concentrate being 8.8 meq/g.

2.3. Termination of Enzymatic Hydrolysis

Termination of enzymatic activity after defined incubation periods or on achievement of a desired %DH was performed in three different ways.

2.3.1. Acid Inactivation and Determination of Residual Enzymatic Activity Effect

Hydrolysis was terminated by adding 1M HCl to achieve pH < 3. Individual samples of 20 mL were readjusted to different pH values (3.0, 6.5, 7.5, 8.5) by addition of 1M NaOH and left overnight at room temperature.

2.3.2. Water Bath Heating

Glass beakers (dimension: d = 4, h = 5 cm) filled with 20 mL of hydrolysate sample were set at different pH values (4.5, 6.5, 7.5, 8.5) and heated in a water bath at 85 °C for 20 min. In order to achieve uniform temperature, the process was undertaken using a magnetic stirrer plate with Telemodul external control set at 200 rpm (Variomag® Telesystem, Thermo-Fisher, Waltham, MA, USA).

2.3.3. Pilot-Scale Heat Exchanger Heating

In order to achieve more rapid heating, a Microthermics UHT/HTSTL lab-25 EHVH (Raleigh, NC, USA) continuous flow heater was used. The equipment was configured to accomplish heating in two successive phases: step 1 raised the temperature of the protein hydrolysate to 60 °C; step 2 that achieved the final temperature of 85 °C with an extended holding time of 5 min, hereafter referred to as HTST-EHT. Samples (3 L) of hydrolysates were prepared and adjusted to different pH values (6.5, 7.0, 7.5, 8.0, 8.5). To avoid contamination or dilution, RO water was pumped through the process between sample runs and the initial and final 500 mL of eluate was excluded from sample collection.

Residual enzymatic activity was measured indirectly by monitoring the total protein peak on the chromatograms following analysis by high-performance liquid chromatography (HPLC).

2.4. Protein Profile Analysis

2.4.1. High-Performance Liquid Chromatography (HPLC)

To study the evolution of the hydrolysis over time, the samples were analysed by reverse-phase (RP) HPLC using an Agilent 1200 series chromatograph (Santa Clara, CA, USA) fitted with a quaternary pump, autosampler and DAD multiple wavelength detector. Measurements were performed at 214 and 280 nm. Analysis was performed using both reverse-phase (RP) and size exclusion analysis (SEC). The RP-HPLC column, ZORBAX stableBond, 300 C-18, 4.6 × 150 mm, 5 μm, was fitted with a guard column and guard carriage (ZCG) (Agilent). The mobile phases used: Milli-Q water (Mobile phase A) and acetonitrile (Mobile phase B) (HPLC-grade ≥99.9%, Sigma-Aldrich, Hamburg, Germany) both containing 0.1% trifluoroacetic acid (TFA) (HPLC-grade provided by Sigma-Aldrich/Merck, Arklow, Ireland,). The RP-HPLC gradient used for the analysis is described in Table 1 with an equilibration (post) time of 4 min in between samples. Calibration curves were produced for individual whey proteins (β-LG, α-LA, BSA, lactoferrin) as well as WPI using known sample concentrations (range from 2.0 to 0.06 mg/mL) prepared from their respective standards (Sigma-Aldrich).

Table 1. High-performance liquid chromatography-reverse-phase solvent gradients used at various time points for protein-peptide separation and quantification.

t (min)	% Acetonitrile
0	3.0
20	38.4
26	38.4
28	44.0
31	52.0
33	90.0
37	90.0
38	3.0

Hydrolysate samples were prepared at a concentration of 2 mg/mL by dilution in MQ water. The flow rate used was 1 mL/min with a column temperature of 35 °C, and 8 μL of sample was injected.

Molecular size distribution profiles were obtained using a size exclusion column (SEC) TSKgel G2000WXL, 7.8 × 300 mm (Tosoh Corporation, Tokyo, Japan) that was protected by a TSKgel SWXL (6 × 40 mm, 7 μm) guard column. The mobile phase in this case was a mixture of Milli-Q water (70%) and acetonitrile (30%) with an 0.1% TFA additive. A flow rate of 0.5 mL/min was maintained over a 40-min period. The sample injection volume was 20 μL at 2 mg/mL concentration. The column was calibrated with the following standards: BSA (66 kDa), a-LA (14 kDa), aprotinin (6.5 kDa), bacitracin (1.42 kDa), polypeptide (leu-Trp-Met-Arg) 604 Da and dipeptide Try-Glu (310 Da).

2.4.2. Electrophoresis

Sodium dodecyl sulphate-polyacrylamide gel electrophoresis (SDS-PAGE) was performed using NuPAGE Novex bis-Tris 12-well precast gels (Invitrogen, Life Technologies Corp., Carlsbad, CA, USA), containing 4–12% polyacrylamide, prepared according to the manufacturer's instructions under reducing and non-reducing conditions in order to study the possible aggregation of proteins using MES (2-(N-Morpholino) ethanesulfonic acid hemisodium salt) buffer. Sample volumes (10 μL) were injected into each well to achieve a protein loading of 0.65 mg/mL. A fixing solution (50% methanol and 10% acetic acid in *v*/*v*) was applied to the gels for 2 h before staining with the commercial staining solution SimplyBlue^{TM} (Thermo Scientific, Vilnius, Lithuania). Thermo Scientific's PageRuler Unstained Low Range Protein Ladder, with its mixture of purified proteins and peptides in the size range 3.4 to 100 kDa, was used as molecular size markers.

3. Results and Discussion

All the supplementary material (Figures S1–S4 and Tables S1–S2) attached contain the data to support the figures and tables included and described in the text.

3.1. Enzymatic Hydrolysis

As described in previous work, trypsin selectively hydrolyses β-LG under specific temperature and pH conditions. In the present study, trypsin hydrolysed 100% and <33% of β-LG and α-LA (Table 1), respectively, after 3 h hydrolysis at an E:S of 1:10, which was consistent with the findings of Cheison et al. [46]. No other E:S ratio evaluated achieved 100% hydrolysis of β-LG, irrespective of the incubation times studied; however, ~90% hydrolysis could be achieved using an E:S of 1:30 and 1:50 within a 4-h incubation period. Extending the incubation period to 6 h using an E:S of 1:50 only marginally increased β-LG hydrolysis >90%.

Lowering E:S from 1:10 to 1:30 reduced the extent of hydrolysis of β-LG and α-LA to 86.5 ± 1.06% and 11.83 ± 0.9%, respectively, during a 3-h incubation. A further decrease in E:S to 1:50 marginally reduced the extent of β-LG hydrolysis to 81.04 ± 0.81%, but with a beneficial effect in terms of limiting the extent of α-LA hydrolysis, which was reduced to <5%. An E:S of 1:100 during a similar 3-h incubation period resulted in 53.15 ± 2.87% hydrolysis of β-LG, and this value could be extended further to 77.54 ± 1.01% during a 7-h incubation. While 86% hydrolysis of β-LG may be achieved during a shorter duration (1 h) of incubation using an E:S of 1:10, its more intense hydrolytic activity resulted in greater hydrolysis of α-LA (25%).

The evolution of hydrolysis at all E:S studied (based on %DH) and the change in composition of whey protein is represented in Figure 1 and Table 2, respectively. Within the DH range 4.5–6.5%, trypsin hydrolysed β-LG to >90% while α-LA remained almost unhydrolysed (<10%), thus indicating key parameters that could be employed for selective recovery of the latter fraction in a near-intact form. Above DH 6.5%, the extent of hydrolysis of α-LA increased.

Thus, the protein hydrolysis protocols chosen for further experimental studies involved an E:S 1:50 and an incubation time of 3 h based on targeted protein hydrolysis levels of ~5% α-LA and ~80% β-LG. These conditions should equate to an incubation time of moderate duration and a relatively low enzyme cost.

Table 2. Comparison of overall degree of hydrolysis (%DH) and extent of hydrolysis (%) of β-lactoglobulin and α-lactalbumin at different enzyme: Substrate ratios (E:S 1:10, 1:30, 1:50 and 1:100) and incubation times (1 to 7 h) during the targeted hydrolysis of whey protein isolate with trypsin during incubation at 25 °C and pH 8.5. Mean values and standard deviation ($n = 3$) were derived from the results of triplicate trials.

	E:S 1:10			E:S 1:30			E:S 1:50			E:S 1:100		
Time (h)	%DH	α-LA (%)	β-LG (%)	%DH	α-LA (%)	β-LG (%)	%DH	α-LA (%)	β-LG (%)	%DH	α-LA (%)	β-LG (%)
1	5.71 ± 0.19	26.84 ± 2.61	86.46 ± 1.87	3.93 ± 0.26	7.43 ± 1.83	70.36 ± 3.10	2.82 ± 0.14	0.05 ± 5.79	58.67 ± 1.51	1.30 ± 0.24	2.39 ± 2.93	30.63 ± 1.58
2	7.00 ± 0.22	31.02 ± 2.15	97.49 ± 0.76	5.07 ± 0.32	9.91 ± 3.27	80.63 ± 0.57	4.01 ± 0.13	0.84 ± 2.34	73.72 ± 1.67	2.14 ± 0.32	2.54 ± 2.23	47.34 ± 1.53
3	7.83 ± 0.17	33.18 ± 1.67	100.00	5.79 ± 0.37	11.83 ± 0.93	86.55 ± 1.06	4.64 ± 0.15	1.51 ± 5.20	81.04 ± 0.81	2.78 ± 0.35	3.52 ± 2.53	53.15 ± 2.87
4				6.43 ± 0.38	11.59 ± 0.93	90.54 ± 1.11	5.08 ± 0.20	3.83 ± 6.92	86.19 ± 0.19	3.32 ± 0.36	3.47 ± 4.99	63.90 ± 0.84
5				6.88 ± 0.44	9.82 ± 2.88	93.14 ± 0.40	5.46 ± 0.23	4.11 ± 3.44	88.32 ± 0.44	3.70 ± 0.34	3.00 ± 1.82	69.81 ± 2.19
6				7.34 ± 0.43	11.06 ± 4.43	95.38 ± 0.60	5.88 ± 0.19	4.43 ± 3.47	91.71 ± 0.45	4.08 ± 0.26	3.13 ± 2.28	73.27 ± 1.35
7				7.86 ± 0.48	14.10 ± 3.31	97.22 ± 0.50	6.17 ± 0.20	4.68 ± 2.88	93.85 ± 0.31	4.39 ± 0.23	4.48 ± 3.22	77.54 ± 1.01

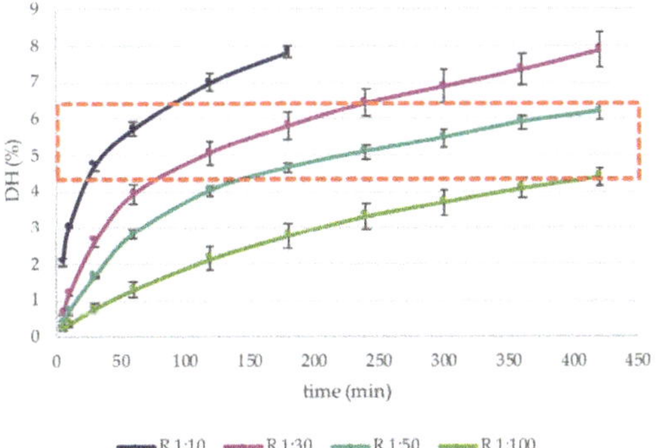

Figure 1. The effect of a targeted hydrolysis protocol (pH 8.5, 25 °C) on the % degree of hydrolysis (%DH) vs. incubation time (min) of whey protein isolate using trypsin at different enzyme:substrate ratios (E:S 1:10, 1:30, 1:50, 1:100). All trials were conducted in triplicate. Values represent mean ± standard deviation.

3.2. Inactivation Assays

3.2.1. Acid Inactivation

There was little evidence of trypsin activity at pH < 3. Enzyme reactivation, however, was observed once the pH was restored stepwise to pH 8.5. The effect after overnight storage of the hydrolysate at room temperature (~20 °C) on the further degradation of α-LA and β-LG, as determined by RP-HPLC, is represented in Figure 2. Between 25 and 30% of the remaining α-LA was hydrolysed within the range pH 5.5–8.5 under these conditions (Figure 2), while those samples adjusted to pH < 3 were not further hydrolysed following overnight storage. It was also notable that α-LA in the samples held within the pH range 5.5–6.5 was preferentially hydrolysed by trypsin relative to β-LG. However, that trend was reversed on pH adjustment to 7.5, resulting in a more extensive hydrolysis of β-LG by trypsin on storage at near-neutral pH conditions. Finally, as expected, trypsin preferentially increased the hydrolysis of β-LG at pH 8.5, as was reported by Cheison et al. [43].

Thus, the termination of enzymatic reactions by lowering the pH is challenging as any subsequent pH alteration can lead to enzyme reactivation and further non-targeted hydrolysis. In any case, the limitation of having to maintain the hydrolysate pH within an acidic pH range restricts further analysis and end-product application.

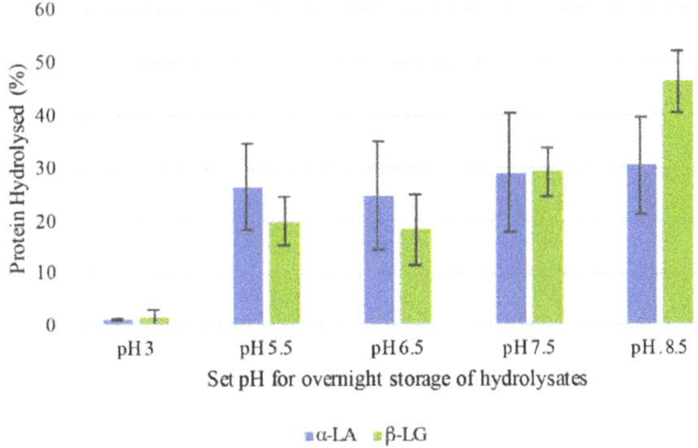

Figure 2. Percentage change in hydrolysis of α-lactalbumin (α-LA) and β-lactoglobulin (β-LG) following overnight storage at room temperature (20 °C) and different pH values (3.0, 5.5, 6.5, 7.5, 8.5) of a tryptic whey protein isolate hydrolysate, analysed in triplicate.

3.2.2. Heat Inactivation: Water Bath

At the laboratory scale, a water bath is the most commonly used method for enzyme inactivation. In order to reduce the heating time, it is necessary to increase the temperature and minimise the sample volume while stirring. However, laboratory heating proved to be ineffective at controlling the trypsin hydrolysis of α-LA during the inactivation process, according to the data obtained from the protein fraction in the RP-HPLC chromatograms at all pH values (Figure 3). Figure 3 also shows that trypsin was still selectively hydrolysing β-LG during heat treatment in the pH range 6.5–8.5, thus resulting in an increased number of peptides produced, although the enzyme was inactivated by the end of the heating. In the case of the samples at pH 7.5 and 8.5, both α-LA and β-LG were completely hydrolysed, thus rendering it impossible to recover α-LA. At pH 6.5 the hydrolytic effect was less pronounced, although much of the intact α-LA and β-LG had disappeared. Just as in the case of the sample held at pH 4.5, the remaining β-LG was not hydrolysed, yet α-LA was not detected.

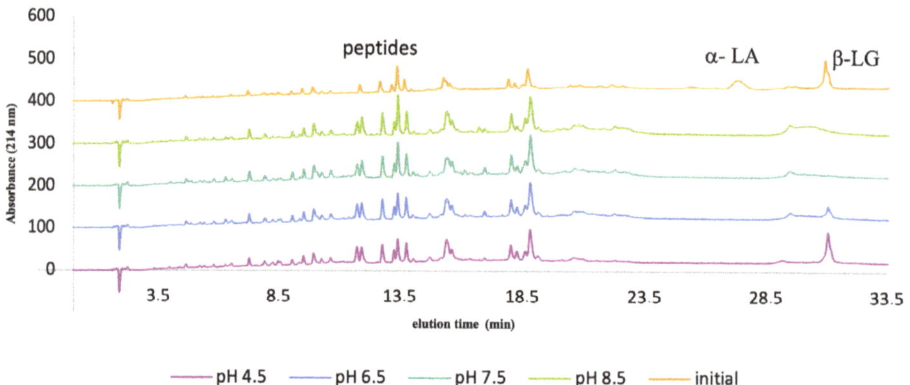

Figure 3. Reverse-phase high-performance liquid chromatography profiles of a tryptic hydrolysate of whey protein isolate at different pH values (4.5, 6.5, 7.5, and 8.5) following water bath heating to 85 °C for 20 min for the purpose of trypsin inactivation. α-LA(α-lactalbumin) β-LG (β-lactoglobulin).

This loss of enzyme selectivity, together with continuing hydrolysis, occurred during heat inactivation of trypsin at a relatively high denaturation temperature of 85 °C for 20 min. In order to attain the target inactivation temperature of 85 °C during heating from 25 °C, it was necessary during the ramping-up period to traverse the region of optimum temperature for trypsin activity, i.e., between 3 °C and 55 °C. As a result, the preferential enzyme selectivity observed at 25 °C was lost during this heat inactivation treatment, thus negating the opportunity to recover α-LA.

3.2.3. Heat Inactivation: Heat Exchanger

In order to improve the thermal denaturation kinetics of trypsin in a whey protein hydrolysate medium, a Microthermics® continuous HTST-EHT heat exchanger was used to rapidly increase the temperature during the heating step, thus minimising the possibility of continuing hydrolysis during transition through the enzyme's operating temperature range. Figure 4 shows the effect of HTST-EHT heat treatment on samples treated at different pH values (6.5, 7, 7.5, 8, 8.5) on the molecular mass profiles when analysed by SEC-HPLC. The results were plotted in conjunction with the elution time of the protein standards: α-LA (14 kDa), β-LG (36 kDa) and BSA (63 kDa) (Figure 4). Surprisingly, there was little evidence of either of these whey protein fractions in the chromatograms of the HTST-EHT heat-inactivated hydrolysates, irrespective of the heating pH used. However, aggregation was evident in all HTST-EHT-heated hydrolysate samples. These aggregates (>66 kDa) are considered to be as a result of the high temperature of 85 °C used to inactivate trypsin, during which protein/peptide aggregate complexes are formed by non-covalent interactions and disulphide aggregation. This development obstructed further quantification and detailed study of the targeted hydrolysates [46,47].

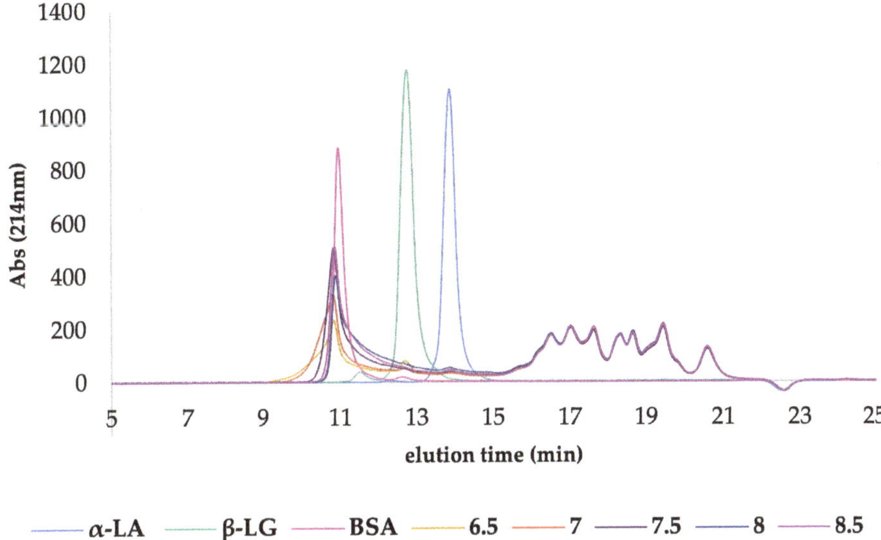

Figure 4. High-performance liquid chromatography-size exclusion (HPLC-SEC) profiles, representing the effect of continuous heat and holding (HTST-EHT) treatment using a heat exchanger on selectively hydrolysed whey protein isolate samples held at different pH values. Protein standards: α-lactalbumin (α-LA), β-lactoglobulin (β-LG), and bovine serum albumin (BSA).

SDS-PAGE under reducing and non-reducing conditions was also used to analyse the effect of heat treatment on the protein hydrolysate. Under reducing conditions (Figure 5A), the aggregates were not evident. At pH 6.5, 7.0 and 7.5 the intensity of the bands of β-LG and α-LA was lower, indicating that both proteins underwent some hydrolysis consistent with the optimum pH for trypsin activity and, thus, affecting the selectivity for α-LA. In the case of samples at pH 8.0 (Lane 5) and 8.5 (Lane 6), both proteins remained intact, confirming the achievement of the desired effect of the selective hydrolysis protocol. Under non-reducing conditions (Figure 5B), it was not possible to detect any β-LG or α-LA bands in lanes 2–6, representing the pH range 6.5–8.5, although there was a faint band representing BSA in lanes 4–7, which coincided with a smaller variation in pH 7.5–8.5. However, under reducing conditions (Figure 5A), bands representing β-LG and α-LA are clearly evident in the pH 8 and 8.5 (lanes 5 and 6) samples, along with BSA (66 kDa). Such a difference generated by SDS-PAGE analysis confirms the occurrence of protein-peptide aggregation produced by HTST-EHT heat treatment, and the reducing conditions necessary to achieve aggregate breakdown.

HTST-EHT continuous thermal processes such as those used industrially would, therefore, be expected to minimise the extent of intact α-LA loss compared with batch heating, e.g., the water bath method as described in this study, due to the lag time taken to reach the inactivation temperature. However, aggregate formation induced during HTST-EHT heat inactivation of a trypsin-containing whey protein hydrolysate makes it practically impossible to recover unhydrolysed α-LA.

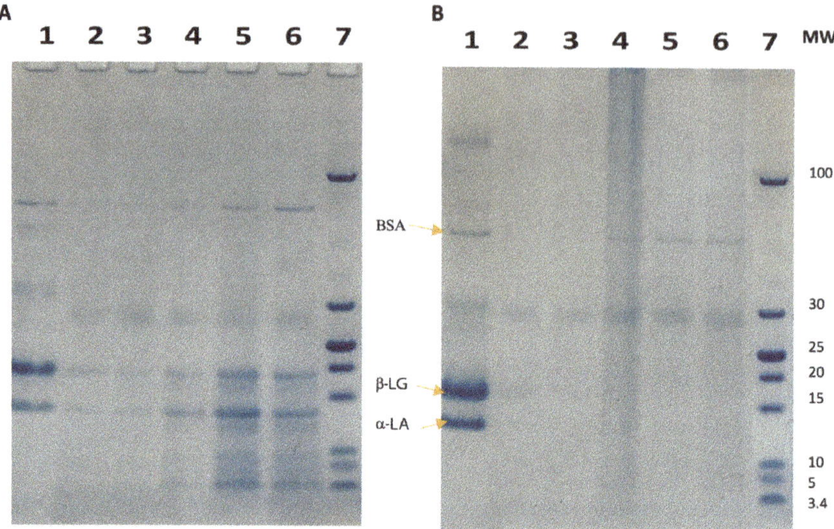

Figure 5. Sodium dodecyl sulphate polyacrylamide electrophoresis (SDS-PAGE) of tryptic hydrolysates of whey protein isolate following heating to 85 °C for 5 min at different pH values, using a heat exchanger, under (**A**) reducing conditions (1) WPI; (2) pH 6.5; (3) pH 7; (4) pH 7.5; (5) pH 8; (6) pH 8.5; (7) molecular weight marker; (**B**) SDS-PAGE under non-reducing conditions, (1) WPI, (2) pH 6.5; (3) pH 7; (4) pH 7.5; (5) pH 8; (6) pH 8.5; (7) molecular weight markers.

4. Conclusions

In the context of future process scale-up and the necessity of controlling the costs of enzyme addition, the optimum E:S was determined to be 1:50 with extension of incubation time from 3 to 7 h in order to selectively achieve 100% hydrolysis of β-LG with minimal degradation of α-LA. Further experiments with an E:S of 1:100 may also be of interest in order to determine the total time needed for 100% hydrolysis of β-LG, while at the same time considering the economic consequences of running hydrolysis over a protracted period. All traditional methods employed to terminate enzymatic activity proved effective, but had a considerable negative impact on the target protein, i.e., α-LA. Acidification to <pH 3 limits possible further application of the hydrolysate, as all functionality and formulation could not be conducted at such an acidic pH. Heat treatment by means of continuous HTST-EHT heating was effective at limiting further protein hydrolysis, with trypsin selectivity affected by the pH—at neutral pH conditions, the selectivity of hydrolysis was compromised and intact α-LA began to be digested, while selectivity for maintaining intact α-LA was not affected at a higher pH (pH > 8). However, heat-induced protein/peptide aggregate formation makes it practically impossible to recover unhydrolysed α-LA intact. For these reasons, continuous HTST-EHT heating was identified as unsuitable for terminating tryptic-led WPI hydrolysis, especially when the process is aimed at limiting the breakdown of intact α-LA. Further studies are required to determine whether the aggregates formed may be broken down subsequently in order to facilitate the recovery of intact α-LA. Other processing options should be investigated for managing and controlling enzyme activity in order to facilitate the downstream recovery of intact α-LA [48–50].

Supplementary Materials: The following are available online at http://www.mdpi.com/2304-8158/8/9/367/s1, Figure S1: DH different ratios, Figure S2: Stopped acid, Figure S3: Water bath, Figure S4: SEC-HPLC, Table S1: RP-HPLC gradient, Table S2: HPLC-RP representation at different ratios.

Author Contributions: L.S. performed the experimental work, analysed data and drafted the manuscript, E.M. supervised daily laboratory routines, R.J.F. provided academic guidance, P.K. conceived the initial project objectives, directed the research study and edited the manuscript throughout drafting

Funding: Work described herein was supported by Enterprise Ireland-funded 'Food for Health Ireland' (FHI) project under Grant Number TC20130001.

Conflicts of Interest: Authors declare no conflict of interest.

References

1. Lahl, W.J.; Braun, S.D. Enzymatic production of protein hydrolysates for food use. *Food Technol.* **1994**, *48*, 68–71.
2. Woiciechowski, A.L.; Nitsche, S.; Pandey, A.; Soccol, C.R. Acid and enzymatic hydrolysis to recover reducing sugars from cassava bagasse: An economic study. *Braz. Arch. Biol. Technol.* **2002**, *45*, 393–400. [CrossRef]
3. Gulati, S.L.; Gaur, A.C. Enzymatic hydrolysis of cellulose from agricultural and industrial wastes. *Biol. Wastes* **1988**, *25*, 309–313. [CrossRef]
4. Guellil, A.; Boualam, M.; Quiquampoix, H.; Ginestet, P.; Audic, J.M.; Block, J.C. Hydrolysis of wastewater colloidal organic matter by extracellular enzymes extracted from activated sludge flocs. *Water Sci. Technol.* **2001**, *43*, 33–40. [CrossRef] [PubMed]
5. Moon, H.C.; Song, I.S. Enzymatic hydrolysis of foodwaste and methane production using UASB bioreactor. *Int. J. Gr. Energy* **2011**, *8*, 361–371. [CrossRef]
6. Mitidieri, S.; Souza Martinelli, A.H.; Schrank, A.; Vainstein, M.H. Enzymatic detergent formulation containing amylase from Aspergillus niger: A comparative study with commercial detergent formulations. *Bioresour. Technol.* **2006**, *97*, 1217–1224. [CrossRef]
7. McCarthy, A.; O'Callaghan, Y.; O'Brien, N. Protein hydrolysates from agricultural crops—Bioactivity and potential for functional food development. *Agriculture* **2013**, *3*, 112–130. [CrossRef]
8. Adler-Nissen, J. Enzymatic hydrolysis of proteins for increased solubility. *J. Agric. Food Chem.* **1976**, *24*, 1090–1093. [CrossRef]
9. Kong, X.; Zhou, H.; Qian, H. Enzymatic hydrolysis of wheat gluten by proteases and properties of the resulting hydrolysates. *Food Chem.* **2007**, *102*, 759–763. [CrossRef]
10. Banach, J.C.; Lin, Z.; Lamsal, B.P. Enzymatic modification of milk protein concentrate and characterization of resulting functional properties. *LWT-Food Sci. Technol.* **2013**, *54*, 397–403. [CrossRef]
11. Selamassakul, O.; Laohakunjit, N.; Kerdchoechuen, O.; Ratanakhanokchai, K. A novel multi-biofunctional protein from brown rice hydrolysed by endo/endo-exoproteases. *Food Funct.* **2016**, *7*, 2635–2644. [CrossRef]
12. Purschke, B.; Meinlschmidt, P.; Horn, C.; Rieder, O.; Jäger, H. Improvement of techno-functional properties of edible insect protein from migratory locust by enzymatic hydrolysis. *Eur. Food Res. Technol.* **2018**, *244*, 999–1013. [CrossRef]
13. Ryan, G.; Nongonierma, A.B.; O'Regan, J.; FitzGerald, R.J. Functional properties of bovine milk protein isolate and associated enzymatic hydrolysates. *Int. Dairy J.* **2018**, *81*, 113–121. [CrossRef]
14. He, S.; Franco, C.; Zhang, W. Functions, applications and production of protein hydrolysates from fish processing co-products (FPCP). *Food Res. Int.* **2013**, *50*, 289–297. [CrossRef]
15. Baraldi, I.J.; Giordano, R.L.C.; Zangirolami, T.C. Enzymatic hydrolysis as an environmentally friendly process compared to thermal hydrolysis for instant coffee production. *Braz. J. Chem. Eng.* **2016**, *33*, 763–771. [CrossRef]
16. Clemente, A. Enzymatic protein hydrolysates in human nutrition. *Trends Food Sci. Technol.* **2001**, *11*, 254–262. [CrossRef]
17. Fiocchi, A.; Restani, P.; Bernardini, R.; Lucarelli, S.; Lombardi, G.; Magazzù, G.; Marseglia, G.L.; Pittschieler, K.; Tripodi, S.; Troncone, R.; et al. A hydrolysed rice-based formula is tolerated by children with cow's milk allergy: A multi-centre study. *Clin. Exp. Allergy* **2006**, *36*, 311–316. [CrossRef]
18. Gauthier, S.F.; Pouliot, Y.; Saint-Sauveur, D. Immunomodulatory peptides obtained by the enzymatic hydrolysis of whey proteins. *Int. Dairy J.* **2006**, *16*, 1315–1323. [CrossRef]
19. Sinha, R.; Radha, C.; Prakash, J.; Kaul, P. Whey protein hydrolysate: Functional properties, nutritional quality and utilization in beverage formulation. *Food Chem.* **2007**, *101*, 1484–1491. [CrossRef]
20. Abeyrathne, E.D.N.S.; Lee, H.Y.; Jo, C.; Suh, J.W.; Ahn, D.U. Enzymatic hydrolysis of ovomucoid and the functional properties of its hydrolysates. *Poult. Sci.* **2015**, *94*, 2280–2287. [CrossRef]
21. Li-Chan, E.C.Y. Bioactive peptides and protein hydrolysates: Research trends and challenges for application as nutraceuticals and functional food ingredients. *Curr. Opin. Food Sci.* **2015**, *1*, 28–37. [CrossRef]

22. Dullius, A.; Goettert, M.I.; de Souza, C.F.V. Whey protein hydrolysates as a source of bioactive peptides for functional foods—Biotechnological facilitation of industrial scale-up. *J. Funct. Foods* **2018**, *42*, 58–74. [CrossRef]
23. Nongonierma, A.B.; O Keeffe, M.B.; Fitz Geralde, R.J. Milk protein hydrolysates and bioactive peptides. In *Advanced Dairy Chemistry: Volume 1B: Proteins: Applied Aspects: Fourth Edition*; McSweeney, P.L.H., O'Mahony, J.A., Eds.; Springer: New York, NY, USA, 2016; pp. 417–482.
24. Tsabouri, S.; Douros, K.; Priftis, K.N. Cow's milk allergenicity. *Endocr. Metab. Immune Disord. Drug Targets* **2014**, *14*, 16–26. [CrossRef]
25. Severin, S.; Xia, W.S. Enzymatic hydrolysis of whey proteins by two different proteases and their effect on the functional properties of resulting protein hydrolysates. *J. Food Biochem.* **2006**, *30*, 77–97. [CrossRef]
26. Walmsley, S.J.; Rudnick, P.A.; Liang, Y.; Dong, Q.; Stein, S.E.; Nesvizhskii, A.I. Comprehensive analysis of protein digestion using six trypsins reveals the origin of trypsin as a significant source of variability in proteomics. *J. Proteome Res.* **2013**, *12*, 5666–5680. [CrossRef]
27. Butré, C.I.; Sforza, S.; Gruppen, H.; Wierenga, P.A. Determination of the influence of substrate concentration on enzyme selectivity using whey protein isolate and Bacillus licheniformis protease. *J. Agric. Food Chem.* **2014**, *62*, 10230–10239. [CrossRef]
28. Deng, Y.; Gruppen, H.; Wierenga, P.A. Comparison of protein hydrolysis catalyzed by bovine, porcine, and human trypsins. *J. Agric. Food Chem.* **2018**, *66*, 4219–4232. [CrossRef]
29. Gbogouri, G.A.; Linder, M.; Fanni, J.; Parmentier, M. Influence of hydrolysis degree on the functional properties of salmon byproducts hydrolysates. *J. Food Sci.* **2004**, *69*, C615–C622. [CrossRef]
30. Le Maux, S.; Nongonierma, A.B.; Barre, C.; FitzGerald, R.J. Enzymatic generation of whey protein hydrolysates under pH-controlled and non pH-controlled conditions: Impact on physicochemical and bioactive properties. *Food Chem.* **2016**, *199*, 246–251. [CrossRef]
31. Chae, H.J.; In, M.J.; Kim, M.H. Process development for the enzymatic hydrolysis of food protein: Effects of pre-treatment and post-treatments on degree of hydrolysis and other product characteristics. *Biotechnol. Bioprocess Eng.* **1998**, *3*, 35–39. [CrossRef]
32. O'Loughlin, I.B.; Murray, B.A.; Kelly, P.M.; Fitzgerald, R.J.; Brodkorb, A. Enzymatic hydrolysis of heat-induced aggregates of whey protein isolate. *J. Agric. Food Chem.* **2012**, *60*, 4895–4904. [CrossRef]
33. Guan, H.; Diao, X.; Jiang, F.; Han, J.; Kong, B. The enzymatic hydrolysis of soy protein isolate by Corolase PP under high hydrostatic pressure and its effect on bioactivity and characteristics of hydrolysates. *Food Chem.* **2018**, *245*, 89–96. [CrossRef]
34. Alting, A.C.; Meijer, R.J.G.M.; van Beresteijn, E.C.H. Selective hydrolysis of milk proteins to facilitate the elimination of the ABBOS epitope of bovine serum albumin and other immunoreactive epitopes. *J. Food Prot.* **2016**, *61*, 1007–1012. [CrossRef]
35. Martinez, K.D.; Baeza, R.I.; Millán, F.; Pilosof, A.M.R. Effect of limited hydrolysis of sunflower protein on the interactions with polysaccharides in foams. *Food Hydrocoll.* **2005**, *19*, 361–369. [CrossRef]
36. Avramenko, N.A.; Low, N.H.; Nickerson, M.T. The effects of limited enzymatic hydrolysis on the physicochemical and emulsifying properties of a lentil protein isolate. *Food Res. Int.* **2013**, *51*, 162–169. [CrossRef]
37. Cucu, T.; Platteau, C.; Taverniers, I.; Devreese, B.; De Loose, M.; De Meulenaer, B. Effect of partial hydrolysis on the hazelnut and soybean protein detectability by ELISA. *Food Control* **2013**, *30*, 497–503. [CrossRef]
38. Nieto-Nieto, T.V.; Wang, Y.X.; Ozimek, L.; Chen, L. Effects of partial hydrolysis on structure and gelling properties of oat globular proteins. *Food Res. Int.* **2014**, *55*, 418–425. [CrossRef]
39. Xu, X.; Liu, W.; Liu, C.; Luo, L.; Chen, J.; Luo, S.; McClements, D.J.; Wu, L. Effect of limited enzymatic hydrolysis on structure and emulsifying properties of rice glutelin. *Food Hydrocoll.* **2016**, *61*, 251–260. [CrossRef]
40. Butré, C.I.; Sforza, S.; Wierenga, P.A.; Gruppen, H. Determination of the influence of the pH of hydrolysis on enzyme selectivity of Bacillus licheniformis protease towards whey protein isolate. *Int. Dairy J.* **2015**, *44*, 44–53. [CrossRef]
41. Permyakov, E.A.; Berliner, L.J. α-Lactalbumin: Structure and function. *FEBS Lett.* **2000**, *473*, 269–274. [CrossRef]
42. Layman, D.K.; Lönnerdal, B.; Fernstrom, J.D. Applications for α-lactalbumin in human nutrition. *Nutr. Rev.* **2018**, *76*, 444–460. [CrossRef]

43. Cheison, S.C.; Leeb, E.; Toro-Sierra, J.; Kulozik, U. Influence of hydrolysis temperature and pH on the selective hydrolysis of whey proteins by trypsin and potential recovery of native alpha-lactalbumin. *Int. Dairy J.* **2011**, *21*, 166–171. [CrossRef]
44. Chelulei Cheison, S.; Brand, J.; Leeb, E.; Kulozik, U. Analysis of the effect of temperature changes combined with different alkaline pH on the β-lactoglobulin trypsin hydrolysis pattern using MALDI-TOF-MS/MS. *J. Agric. Food Chem.* **2011**, *59*, 1572–1581. [CrossRef]
45. Lisak, K.; Toro-Sierra, J.; Kulozik, U.; Bozanic, R.; Cheison, S.C. Chymotrypsin selectively digests β-lactoglobulin in whey protein isolate away from enzyme optimal conditions: Potential for native α lactalbumin purification. *J. Dairy Res.* **2013**, *80*, 14–20. [CrossRef]
46. Wijayanti, H.B.; Bansal, N.; Deeth, H.C. Stability of whey proteins during thermal processing: A Review. *Compr. Rev. Food Sci. Food Saf.* **2014**, *13*, 1235–1251. [CrossRef]
47. Le Maux, S.; Nongonierma, A.B.; Lardeux, C.; FitzGerald, R.J. Impact of enzyme inactivation conditions during the generation of whey protein hydrolysates on their physicochemical and bioactive properties. *Int. J. Food Sci. Technol.* **2018**, *53*, 219–227. [CrossRef]
48. Guadix, A.; Camacho, F.; Guadix, E.M. Production of whey protein hydrolysates with reduced allergenicity in a stable membrane reactor. *J. Food Eng.* **2006**, *72*, 398–405. [CrossRef]
49. Cheison, S.C.; Wang, Z.; Xu, S.Y. Use of response surface methodology to optimise the hydrolysis of whey protein isolate in a tangential flow filter membrane reactor. *J. Food Eng.* **2007**, *80*, 1134–1145. [CrossRef]
50. Cheison, S.C.; Wang, Z.; Xu, S.Y. Hydrolysis of whey protein isolate in a tangential flow filter membrane reactor. II. Characterisation for the fate of the enzyme by multivariate data analysis. *J. Membr. Sci.* **2006**, *286*, 322–332. [CrossRef]

Article

The Effect of Calcium, Citrate, and Urea on the Stability of Ultra-High Temperature Treated Milk: A Full Factorial Designed Study

Maria A. Karlsson [1,*], Åse Lundh [1], Fredrik Innings [2], Annika Höjer [3], Malin Wikström [3] and Maud Langton [1]

1 Department of Molecular Sciences, Swedish University of Agricultural Sciences, P.O. 7015, 75007 Uppsala, Sweden; ase.lundh@slu.se (Å.L.); maud.langton@slu.se (M.L.)
2 Tetra Pak Processing Systems AB, Ruben Rausings Gata, 22186 Lund, Sweden; fredrik.innings@tetrapak.com
3 Norrmejerier Ek. Förening, Mejerivägen 2, 90622 Umeå, Sweden; annika.hojer@norrmejerier.se (A.H.); malin.wikstrom@norrmejerier.se (M.W.)
* Correspondence: maria.a.karlsson@slu.se; Tel.: +46-739-544778

Received: 31 July 2019; Accepted: 12 September 2019; Published: 17 September 2019

Abstract: The composition of raw milk is important for the stability of dairy products with a long shelf-life. Based on known historical changes in raw milk composition, the aim of this study was to get a better understanding of how possible future variations in milk composition may affect the stability of dairy products. The effects of elevated calcium, citrate, and urea levels on the stability of ultra-high temperature (UHT) treated milk stored for 52 weeks at 4, 20, 30, and 37 °C were investigated by a two-level full factorial designed study with fat separation, fat adhesion, sedimentation, color, pH, ethanol stability, and heat coagulation time as response variables. The results showed that elevated level of calcium lowered the pH, resulting in sedimentation and significantly decreased stability. Elevated level of citrate was associated with color, but the stability was not improved compared to the reference UHT milk. Elevated levels of urea or interaction terms had little effect on the stability of UHT milk. Storage conditions significantly affected the stability. In conclusion, to continue produce dairy products with high stability, the dairy industry should make sure the calcium content of raw milk is not too high and that storage of the final product is appropriate.

Keywords: UHT milk; interaction effects; calcium; citrate; urea; storage temperature; storage time; shelf-life

---

1. Introduction

The composition and properties of raw milk are important for the manufacture of ultra-high temperature (UHT) treated milk to ensure a final product with high stability during storage, meeting the expected shelf-life. In the UHT process, milk is subject to high temperatures, above 135 °C for a few seconds, a treatment which may induce changes in the stability of the milk. Low stability can lead to fat separation, sedimentation, gelation, and browning [1]. Parameters considered to have most influence on the stability of UHT milk include protein composition, ionic calcium content, pH and parameters affecting all of these [2].

Calcium is important for the internal structure and stability of the casein micelle, both via the colloidal calcium phosphate associated to casein molecules and by the formation of calcium bridges between negatively charged residues of the caseins [3,4]. In milk serum, calcium can form complexes with other agents, or exists as free ions. Total calcium content in milk has been reported at average levels of 26–32 mM [5]. The calcium equilibria between the colloidal and serum phase is affected by pH, and a reduction in pH will increase the calcium ion concentration in the milk serum [3,6]. Also, temperature affects the calcium equilibria. When milk is stored at refrigerated temperatures, calcium phosphate will dissolve, whereas at higher temperatures, calcium phosphate will migrate into the micelle [7]. Previous studies comparing different calcium salts, i.e., calcium chloride, calcium lactate, and calcium gluconate, found that the effect on pH, heat coagulation time, and ethanol stability varied depending on the type of salt used [8,9].

Citrate, mainly present in milk serum, plays an important role in the mineral equilibria of milk and is part of a buffering system between calcium and hydrogen ions. Citrate improves the stability of the milk by forming soluble complexes with calcium, preventing precipitation of calcium phosphate [10]. High levels of citrate have been shown to increase calcium levels in milk serum and consequently a depletion of the colloidal calcium in the casein micelles [4]. The average value of citrate in milk is 7–11 mM [5], varying significantly with feed and stage of lactation [11,12].

Milk urea is known to vary with stage of lactation, season, and feed [13–15]. A higher concentration of urea acts as a stabilizing agent on the micelle [16]. It has been suggested that elevated levels of urea prevent crosslinking and aggregation of proteins, and also keep the pH stable [2]. It is suggested that at pH ≥6.8 urea, degraded to ammonia, has a buffering and stabilizing effect on milk protein concentrate suspensions, observed as longer heat coagulation time [17]. An average urea level of 4.7 mM in Swedish milk was reported by Lindmark-Månsson [18].

The stability of the milk can be evaluated by sensory attributes, including fat separation, fat adhesion to the packaging material, sediment formation, and color, varying with storage time and temperature [19]. During storage of UHT milk, fat globules float to the top, resulting in fat separation and fat adhesion [1]. Fat separation is closely correlated to and will increase with fat content, storage temperature, and milk fat globule size [20,21]. In UHT products, sediment formation is also a well-known problem. It is suggested that sediment consists of aggregates of proteins or protein particles of various sizes [22]. Casein micelles can, by the influence of gravity, sediment under their own weight [23]. Aggregation of micelles will further increase the rate of sedimentation [23]. Sediment formation has also been shown to increase with storage temperature [1,22]. During storage change in the color of UHT milk is mainly affected by the Maillard reaction, resulting in the formation of brown pigments in milk stored at warm temperatures [24].

To predict the stability of the resulting UHT milk, various tests have been used to assess the suitability of the raw milk for UHT processing, e.g. pH, ethanol stability, and heat coagulation time (HCT). Milk has a natural pH around 6.7, and lowering the pH will reduce the net charge on proteins and promote protein-protein interactions [25]. According to manufacturers of UHT processing equipment, raw milk should have a pH above 6.65 to be suitable for UHT processing [26]. Another commonly applied method for detection of poor quality milk is by evaluating its ethanol stability, which is a simple, cheap, and quick pass-fail-test [27]. It is recommended that milk for UHT processing should have an ethanol stability of 74% or higher [28]. Tsioulpas et al. [3] found that milk samples with low free calcium ion concentrations had consistently high ethanol stability, and it has been suggested that factors reducing the negative net charge of the casein micelle may reduce ethanol stability. The heat stability of milk is another commonly used predictive test of raw milk aimed at UHT processing. It is determined by the HCT, i.e., the time it takes for milk to visually coagulate when heated to temperatures above 100 °C [29]. Raw milk with poor heat stability is not suitable for UHT processing, due to increased fouling and sediment formation [2]. The heat stability of milk has been shown to be highly dependent on pH, with maximum heat stability around pH 6.65–6.70 [8].

The composition of milk delivered to Swedish dairies was previously investigated in 2001, and again in 2009 [18,30]. Comparing the composition of raw milk from 2009 with the composition reported in earlier studies, several changes that can be expected to have a negative impact on the stability of UHT milk were observed, for example, the calcium content had increased by 5%, whereas citrate and urea had decreased by 20% and 7%, respectively. Therefore, to get a better understanding of future scenarios, a study was designed to evaluate the impact of levels of calcium, citrate, and urea and their interaction effects on the stability of UHT milk during one year of storage at 4, 20, 30, and 37 °C. The study was set up as a full factorial experimental design, using fat separation, fat adhesion, sediment formation, color, pH, ethanol stability, and heat coagulation time as indicators of stability during storage.

2. Materials and Methods

2.1. Design, Sample Preparation, Handling, and Storage

The experiment was set up as a 2^3 full factorial design. In total nine batches of milk were prepared, at two production occasions, with elevated levels of calcium, citrate, and urea, including two unmodified reference batches (Table 1). Milk containing 20% higher concentrations of calcium, citrate, and urea were prepared based on previously reported average levels of 32, 9, and 4.7 mM of calcium, citrate, and urea, respectively [5,18]. Calcium ($CaCl_2{\cdot}2H_2O$, VWR Chemicals, Leuwen, Belgium), citrate ($Na_3C_6H_5O_7{\cdot}2H_2O$, VWR Chemicals, Leuwen, Belgium), and urea ($CO(NH_2)^2$, Alfa Aesar, Karlsruhe, Germany) were dissolved in water and added to batches of 200 L pasteurized, standardized (1.5% fat), and homogenized milk to obtain calculated final concentrations of 38, 11, and 5.6 mM, in the milk. The modified milk was stored refrigerated overnight and subjected to UHT processing the following day. UHT treatment was conducted at the Tetra Pak Product Development Centre in Lund, Sweden, using upstream homogenization (150 + 30 bar), followed by indirect tubular heat exchangers at 137 °C for 4 seconds, and aseptically packed in 250 mL Tetra Brik Aseptic (TBA) packages. The UHT milk was transported at ambient temperature to the Swedish University of Agricultural Sciences (SLU), Uppsala, analyzed the day after delivery to SLU, so that the UHT milk was not older than 5 days post-production and thereafter stored at 4, 20, 30, and 37 °C for up to 52 weeks after production. During storage, new packages with milk were opened and analyzed every third week. Before the evaluation, samples were stored at room temperature overnight and analyzed the following morning. All measurements were done at ambient temperature unless otherwise stated.

Table 1. Experimental design of the 2^3 full factorial study. Milk with elevated levels of calcium, citrate, and urea were prepared as below and the samples were subject to ultra-high temperature processing. Samples 1–5 and 6–9 were produced at different occasions, with one reference sample included at each occasion (samples 5 and 6). Abbreviations: +20% = elevated level; - = normal level.

Sample	Calcium	Citrate	Urea
1	+20%	-	-
2	-	+20%	-
3	-	-	+20%
4	+20%	+20%	+20%
5	-	-	-
6	-	-	-
7	+20%	+20%	-
8	+20%	-	+20%
9	-	+20%	+20%

2.2. Fat Separation, Fat Adhesion, Sediment Formation, and Color

Fat separation, fat adhesion, sediment formation, and color were measured as described in Karlsson et al. [19]. Fat separation was defined as the thickness of the cream layer floating on the surface and rated on a four-graded scale, i.e., no visual cream layer, waves of cream, a surface completely covered with fat or lumps/clots of fat. Fat adhesion, defined as the thickness and amount of fat adhering to the inside of the package after the milk was poured out, was compared with reference photos and rated on a scale of 0–4, modified from a protocol by the New Zealand Dairy Industry [31]. The amount of sediment formed at the bottom of the package was visually estimated and compared with reference photos, and graded on a scale 0–100%, where 0 corresponded to no sediment and 100 corresponded to the bottom of the package being completely covered by sediment [32]. Milk with sediment covering more than 45% of the bottom of the package was regarded as not acceptable for consumption. The CIELAB color space was measured with a CM-600d spectrophotometer (Konica Minolta, Shanghai, China). Using this technology, L^* indicates lightness ranging from 0–100, a^* indicates a range from green to red (−60 to +60), and b^* a range from blue to yellow (−60 to +60). Milk with L^* values under 76 was considered not acceptable for consumption.

2.3. pH, Ethanol Stability, and Heat Coagulation Time

The pH, ethanol stability, and HCT were measured as described in Karlsson et al. [33]. In brief, pH was measured using an IoLine electrode (SI Analytics®, Mainz, Germany). Ethanol stability was defined as the highest ethanol concentration added to the sample without causing visual coagulation of the milk when equal volumes of milk and ethanol, at ethanol concentrations ranging between 40 and 100% in 2% increments, were mixed and incubated for 30 min [3]. HCT was defined as the time needed for visual coagulation of 0.5 mL milk in a sealed test tube whilst being rocked at 130 °C [29] using the dedicated equipment from Hettich Benelux (Geldermalsen, Netherlands).

2.4. Statistical Analysis

Partial least squares regression (PLS) analysis was carried out using MODDE 11 software (Umetrics, Umeå, Sweden) with the factors calcium, citrate, and urea and the interaction terms calcium*citrate, calcium*urea, citrate*urea, and calcium*citrate*urea. Responses were fat separation, fat adhesion, sedimentation, color, pH, ethanol stability, and HCT. The first four principal components of the PLS model were fitted. Data were normalized by dividing the coefficients with the standard deviation of their respective response. Considering the known strong impact of storage temperature and storage time on stability, how the factors vary with storage temperature and storage time was also studied in the statistical evaluation.

3. Results and Discussion

3.1. Partial Least Squares Regression

The systematic variation in the first principal component can be explained by calcium's main effect on all factors, positively correlated to sediment formation and fat separation, and negatively correlated to all other factors (Figure 1 and Table 2). Citrate and the interaction between calcium and citrate, explaining the variation in the second principal component, were found to mainly affect the color (Figure 1 and Table 2). Urea and the interaction terms, except calcium and citrate, had no effect on the factors included in this study and were to be found at the center of the PLS plot (Figure 1 and Table 2). The four first principal components in total only explained 42% of the systematic variation and the constants of the regression coefficients were high (Table 2), indicating that additional variables, not included in the PLS model, could explain the variation in the data. For example, as discussed below, it is known that storage time and storage temperature affect the stability of UHT milk.

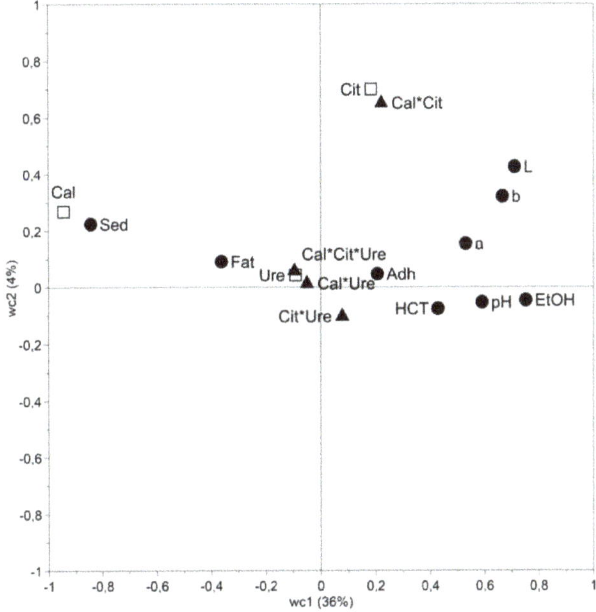

Figure 1. Partial least squares (PLS) for ultra-high temperature treated milk stored for 0-52 weeks at 4, 20, 30, and 37 °C. (□) factors, (▲) interaction terms, and (●) responses. The systematic variation in the data is to 36% and 4%, explained by the first and second principal components, respectively. Abbreviations: a = *a**; Adh = fat adhesion; b = *b**; Cal = calcium; Cit = citrate; EtOH = ethanol stability; Fat = fat separation; HCT = heat coagulation time; L = *L**; Sed = sediment formation.

Table 2. Significant regression coefficients relating to normalized variables for ultra-high temperature treated milk stored for 0–52 weeks at 4, 20, 30, and 37 °C. The size of the coefficient represents the change in response when a factor varies from normal to elevated level. *** $p \leq 0.001$, and ** $p \leq 0.01$.

	Constant	Factors			Interaction Terms			
		Calcium	Citrate	Urea	Cal*Cit	Cal*Urea	Cit*Urea	Cal*Cit*Urea
Fat separation	1.71 ***	0.39 ***						
Fat adhesion	2.16 ***	−0.18 ***	0.13 ***					
Sediment	1.38 ***	0.88 ***			−0.05 ***			
*L**	5.90 ***	−0.61 ***	0.40 ***		0.47 ***			
*a**	−1.12 ***	−0.49 ***	0.23 ***		0.20 ***			
*b**	1.41 ***	−0.59 ***	0.37 ***		0.34 ***			
pH	36.53 ***	−0.61 ***	0.14 ***					
Ethanol stability	4.81 ***	−0.77 ***	0.23 ***					−0.07 **
Heat coagulation time	0.66 ***	−0.44 ***						

3.2. *Effect of Calcium*

In this study, calcium had a significant effect on all responses, and elevated calcium content did, in multiple ways, give a less stable product compared to the reference UHT milk. Elevated calcium levels strongly correlated to sediment formation (Figure 1). Within a week after UHT processing, the entire bottom of the package was covered with sediment (Figure 2) and the product was thereby no longer acceptable for consumption, as the sediment covered >45% of the bottom of the package. Compared to the reference UHT milk, the *L**, *a**, and *b** values were lower for UHT milk with high calcium content, giving a less white, less red, and less yellow product (Figure 3), likely due to fewer light scattering particles. In the same milk, the pH was 6.5 after UHT processing, remaining at this pH during storage at 4 °C, however, it decreased to 6.0 when stored for 52 weeks at 37 °C (Figure 4). Earlier studies have shown a correlation between sediment formation, color, and pH [34] and in agreement with our results, sediment formation was heavy, at a pH below 6.6 [35]. It is assumed that as pH is lowered, the calcium equilibria is shifted, leading to increased ionic calcium levels in serum and less colloidal calcium phosphate in the micelles, hence the casein micelles are destabilized, promoting aggregation of micelles, and sediment formation [36]. Consequently, the reflectance drops, due to a lower number of light scattering particles in milk serum. Also, in studies by Lewis et al. [6], an addition of 4.5 mM calcium chloride, corresponding to approximately 15% supplementation of calcium, resulted in large amounts of sediment at a pH of <6.6. Ramsey and Swartzel [1] and Malmgren et al. [22] found the amount of sediment to strongly depend on an increase of storage time and temperature. In contrast, our results do not clearly show an increase of sediment with increasing storage temperature. As seen in Figure 2, most sediment was formed in UHT milk stored at 4 °C, where the bottom of all packages were fully covered by sediment, corresponding to 100% sediment formation after 52 weeks of storage. In contrast, only 30% sediment was formed in the reference UHT milk when stored at 37 °C for 52 weeks (Figure 2). Our results agree with earlier studies by Wilson, Herreid, and Whitney [37], who also found more sediment formed in milk stored at 4 °C than at 21 and 38 °C. It has been suggested that different mechanisms are attributed to sediment formation at different temperatures [22,38]. The driving mechanism for casein micelle disintegration at low temperatures is the dissociation of β-casein. Whereas in UHT milk stored at high temperatures, sediment consists of large κ-casein depleted micelles [22,38].

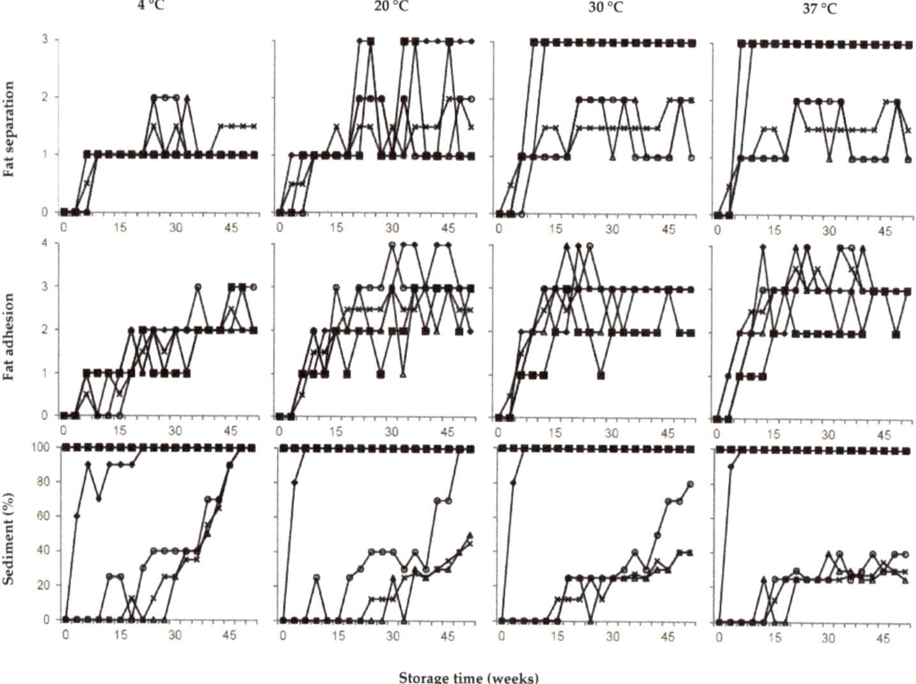

Figure 2. Variation in sediment formation, fat separation, and fat adhesion in ultra-high temperature treated milk for selected samples with elevated levels of (■) calcium, (○) citrate, (Δ) urea, (♦) calcium and citrate, and average values of two batches (×) unmodified reference ultra-high temperature treated milk during storage from 0–52 weeks at 4, 20, 30, and 37 °C.

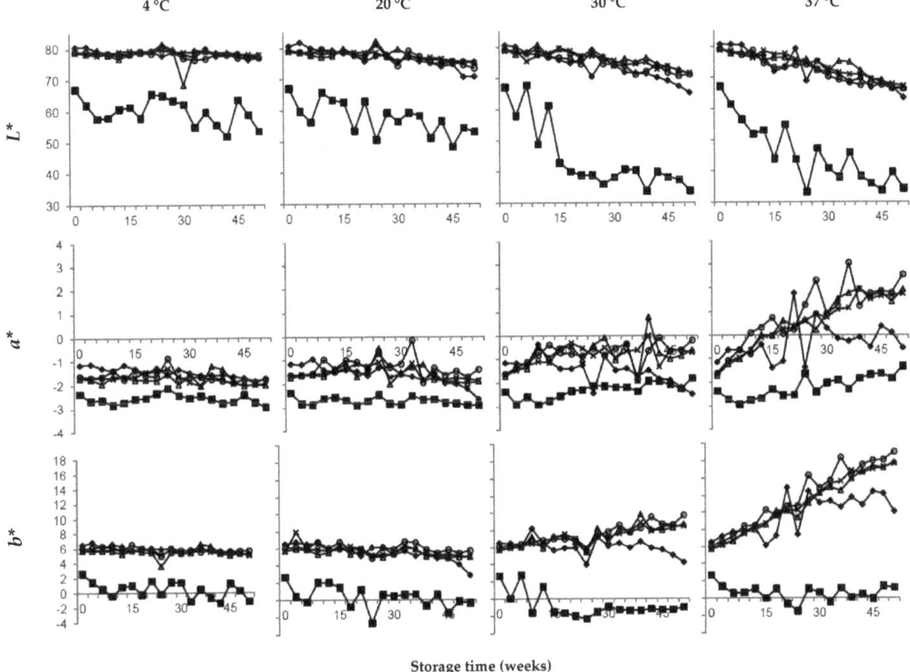

Figure 3. Variation in L^* (lightness), a^* (green-red), and b^* (blue-yellow) in ultra-high temperature treated milk for selected samples with elevated levels of (■) calcium, (○) citrate, (Δ) urea, (♦) calcium and citrate, and average values of two batches (×) unmodified reference ultra-high temperature treated milk during storage from 0–52 weeks at 4, 20, 30, and 37 °C.

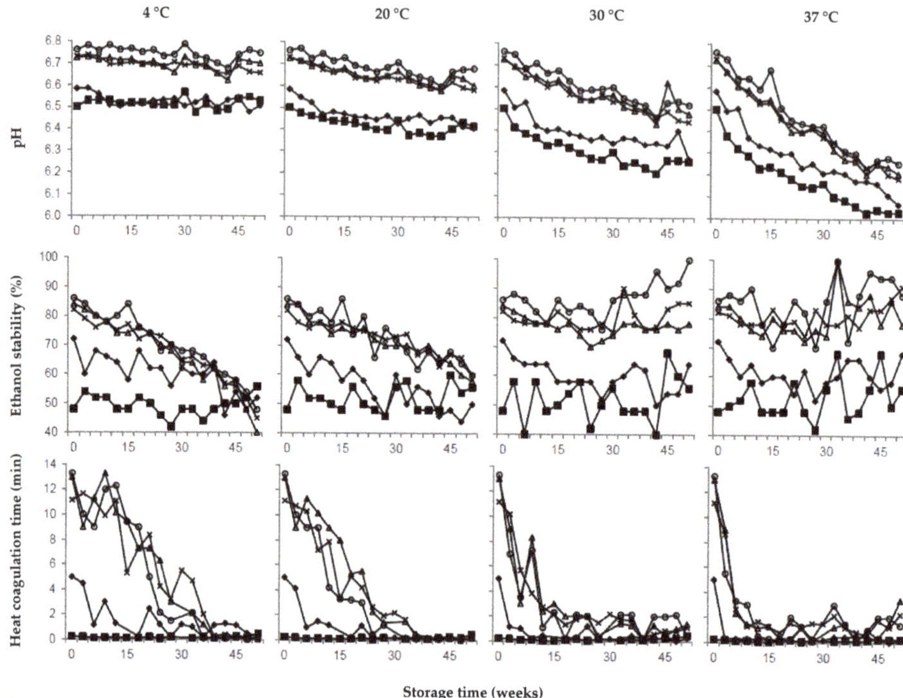

Figure 4. Variation in pH, ethanol stability, and heat coagulation time in ultra-high temperature treated milk for selected samples with elevated levels of (■) calcium, (○) citrate, (Δ) urea, (♦) calcium and citrate, and average values of two batches (×) unmodified reference ultra-high temperature treated milk during storage from 0–52 weeks at 4, 20, 30, and 37 °C.

In UHT milk with elevated calcium levels, ethanol stability was around 50% and HCT was less than a minute, independent of storage temperature and time. In comparison, it has been suggested that milk should have an ethanol stability of >74% to be suitable for UHT production [28]. In agreement with our results, Boumpa et al. [39] found a strong correlation between ethanol stability and ionic calcium. The addition of calcium is known to reduce the negative net charge of the casein micelles, resulting in lower ethanol stability. The effect of elevated levels of calcium on HCT seen in our study corresponds with results by Jeurnink and de Kruif [40], who reported that the HCT for pasteurized milk decreased from 16 to 10 minutes when the calcium content increased by 20%. In milk with high calcium activity, the electrostatic repulsion between micelles will be reduced and the stability will decrease, explaining the instant decline in HCT in UHT milk with elevated calcium content [40,41].

Previous studies have shown that different calcium salts have different effects on the stability of the milk [8,9], and that calcium chloride, which was used in this study, has a large negative impact on the stability of milk. The significant negative effect on the stability of milk by increasing the calcium content by 20% implies that lower levels of calcium chloride or other calcium salts should be used in future studies.

3.3. *Effect of Citrate*

When using a factorial design, it is possible to extract the main effects, comparing all values at high levels and low levels, and thereby increasing the statistical power of the analysis. Thus, comparing the average values of all four samples with elevated citrate content, i.e., also including the samples that in addition to elevated citrate content contained elevated contents of calcium and/or urea, regression coefficients showed that the factor citrate had a significant but small effect on the color of the UHT milk and was also correlated to fat adhesion, pH, and ethanol stability (Table 2). Compared to the reference UHT milk, the one UHT milk sample with elevated citrate content had a similar fat adhesion and color (Figures 2 and 3), a slightly higher pH, and higher ethanol stability (Figure 4), and our results also show a tendency to more sediment formation (Figure 2). To reduce sediment formation in milk with high ionic calcium content, Gaur et al. [36] suggested the addition of citrate, as this increases the pH and chelates calcium. Zadow [34] showed that the addition of as little as 0.3% sodium citrate resulted in less sediment formation. Furthermore, less sediment gave a whiter skim milk, due to the remaining number of light scattering particles. The stabilizing effect of calcium chelators, e.g., citrate, has been shown to depend on calcium activity [42] and the pH [43] of the milk. Previous work has also shown that the addition of 5 mM citrate is required to have an effect on stability [43]. Consequently, in this study, elevating the citrate level by 2 mM did not improve the stability during storage compared to the unmodified reference UHT milk. It is therefore possible that the experimental design of our study, with only 2 levels of citrate, generated results that are somewhat misleading regarding the stabilizing effect of citrate.

3.4. Effect of Urea

In this study, the factor urea is located in the center of the PLS models (Figure 1), meaning that the factor had no effect on responses (Table 2). As seen in Figures 2 and 3, the sensory attributes and, as in Figure 4, the stability during storage, were not improved compared to the reference UHT milk. Milk urea is known to vary with feed, and in a study by Reid et al. [15], cows with high crude protein intake had a significantly higher milk urea concentration compared to intake of a low protein feed. However no difference in heat coagulation time was reported [15]. In agreement with our results, Muir and Sweetsur [44] found that at a pH of 6.6–7.2, urea does not change the mechanism of the coagulation reaction, thus, in this pH region, addition of urea does not lead to a longer heat coagulation time or a higher ethanol stability. Earlier studies indicate that a level of >7 mM of urea is needed to improve stability [44]. In our study, urea was added to a final calculated concentration of 5.6 mM and compared to results from earlier studies, but this was too low to have an effect on the responses.

3.5. Effect of Interaction Terms

The interaction term calcium*citrate had a significant effect on sediment and color (Table 2). In our study, the combination of elevated levels of calcium and citrate had a short delay of few weeks on sediment formation during storage (Figure 2), but could not prevent heavy sediment formation. The actual effect of the calcium*citrate interaction on sediment formation could therefore be questioned. The regression coefficient was significant, but small, −0.05*** (Table 2) and the interaction effect shows almost parallel lines (Figure 5, left), which would indicate a low interaction.

During storage, the L^* values of UHT milk, with the combination of high calcium and citrate content, did not differ from L^* values of the reference UHT milk (Figure 3). And, as seen in the interaction plot (Figure 5, right), elevated levels of calcium in combination with elevated levels of citrate did not give a reduction in lightness (L^*). The interaction effect of calcium and citrate on L^* (Figure 5) shows the same pattern as the interaction effect on a^* and b^* (results not shown), but with lower regression coefficients (Table 2). During storage at 4 and 20 °C, a^* and b^* values of UHT milk with high calcium and citrate content corresponded with values for the reference UHT milk. However, when stored at 30 and 37 °C, the increase in a^* and b^* that took place in the reference UHT milk were less pronounced, and less pigment was formed in UHT milk with elevated calcium and citrate content. This is possibly related to heavy sediment formation, leaving less particles in serum available for the Maillard reaction, hence less change in color during storage. All other interactions (calcium*urea, citrate*urea, and calcium*citrate*urea) had very little effect and are located at the centers of the PLS plots (Figure 1).

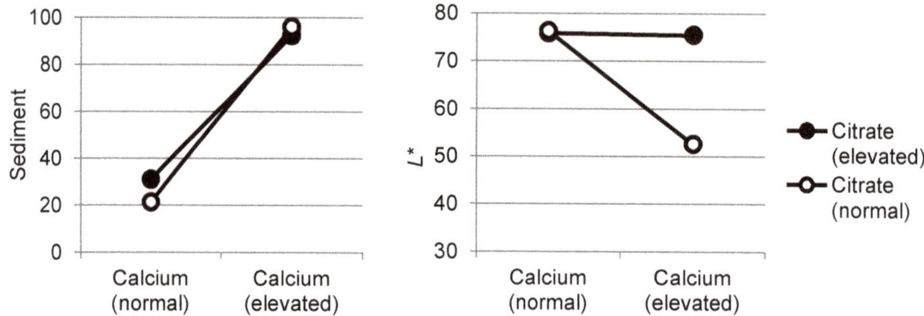

Figure 5. The interaction effects of calcium and citrate, for average values of all samples with normal and elevated levels, on sediment formation and L^* (lightness) for ultra-high temperature treated milk stored for 0–52 weeks at 4, 20, 30, and 37 °C.

3.6. Effect of Storage Temperature and Storage Time

Storage time and storage temperature were not part of the full factorial designed model to evaluate the effects of calcium, citrate, and urea. However, it is know that storage temperature and storage time affect the stability of UHT milk [19]. Our results show that independent of storage temperature, storage time was closely connected to fat separation and fat adhesion, i.e., the longer the storage, the more fat was floating on the surface and adhering to the package (Figure 2). In agreement with Ramsey & Swartzel [1], our study showed that as storage temperature increases, fat separation increases, as explained by Stoke's law. In the reference UHT milk stored at 4 °C, formation of sediment was visible after 24 weeks of storage and thereafter increased linearly, at 52 weeks after production covering the entire bottom of the package (100%) (Figure 2). Sediment formation limited the shelf-life of the UHT milk stored at 4 °C to 40 week, thereafter >45% of the bottom of the package was covered, and the milk was no longer acceptable for consumption. At 20, 30, and 37 °C, at the end of the storage time, the reference milk had sediment covering 40% of the bottom of the package, never reaching an unacceptable level.

The lightness (L^*) of the reference UHT milk stored at 4 °C remained around 80 during the 52 weeks of storage (Figure 3), whereas when stored at 20, 30, and 37 °C, L^* values decreased linearly to 75, 70, and 65, respectively. Changes in L^* can be explained by less light scattering particles in the solution, due to fat separation and sediment formation. The PLS models (Figure 1) showed that the a^* and b^* values were correlated, hence a more red sample would also be more yellow. For the reference UHT milk stored at 4 and 20 °C, a^* and b^* values did not change during storage, whereas milk stored at 30 and 37 °C became considerably more brown (more red and yellow), corresponding to higher a^* and b^* values (Figure 3). At 37 °C, the change in color was most apparent, with a^* and b^* values increasing from −1 to 2 and from 6 to 16, respectively (Figure 3), eventually resulting in brown-colored milk. When UHT milk is stored warm, the Maillard reaction will contribute to the formation of brown pigment [45]. Our results corresponded with findings reported by Gaucher et al. [46] measuring b^* values of around 15 in UHT milk stored at 40 °C for up to 26 weeks.

In the reference UHT milk stored at 4 °C, the pH more or less remained at its initial value of 6.7 throughout the 52 weeks of storage (Figure 4). In contrast, at 37 °C, pH values decreased linearly by 0.5 units during storage from weeks 0 to 52. Our results for changes in pH corresponded with a study by Al-Saadi and Deeth [45], who stored UHT milk for 12 weeks at 5, 20, and 37 °C, and found the pH decreased from 6.6 to 6.5 when stored at 37 °C. Small changes by 0.1 pH have been reported to result in large changes in heat stability [2,40,41]. The Maillard reaction, including the development of formic acid, proceeds faster at a higher storage temperature, explaining the differences in pH between storage temperatures [45], as well as the rapid decrease in HCT during storage at elevated temperatures.

In this study, after UHT treatment, the reference milk had an initial ethanol stability of >80% and an HCT of >10 min (Figure 4), decreasing during storage at 4 °C to around 50%, and when stored at 20 °C, went down to around 60%. In contrast, for UHT milk stored at 30 °C and 37 °C, we observed a tendency for increased ethanol stability during storage. It is known that during cold storage of milk calcium and phosphate, β-casein and other caseins will dissociate from the micelles into the milk serum, due to increased solubility at lower temperature and the weakening of hydrophobic bonds [5,41], resulting in reduced stability of the micelle, possibly explaining the decrease in ethanol stability and HCT during storage at 4 °C (Figure 4). In our study, the decrease in ethanol stability at 4 °C was not correlated to a reduction in pH (Figure 4). Neither was the decrease in pH during storage at 37 °C correlated to lower ethanol stability. In a recent publication, the visual coagulation in milk failing the ethanol stability test was believed to originate from a collapse of the outer hairy κ-casein layer of the casein micelle, causing the micelles to aggregate [25]. However, further studies are needed to fully understand the variation in ethanol stability during storage of UHT milk.

4. Conclusions

Calcium content, and the associated reduction in pH, had a significant negative effect on the stability of UHT milk and were strongly correlated to extensive sediment formation. Addition of citrate mainly affected the color of the UHT milk, and it is suggested that higher concentrations than those used in our study are needed to have a significant improvement on stability. The calcium*citrate interaction had a small but significant effect on sediment and color, whereas all other interactions, and urea, had no effect and thus did not affect the stability. Future research should be careful in designing experiments that involve milk's colloidal stability and its related functionality in products. Measuring just a few levels of the major components that affect the colloidal stability of milk, like calcium, citrate and urea in this study, may exclude the critical combinations of the parameters. Regrettably, the data generated in this study give little new information about possible drivers for the aggregation phenomena behind sedimentation. Still, for the dairy industry to manufacture products of high quality and with high stability, monitoring the natural trends in milk composition, especially variations in calcium content, is important.

Author Contributions: Conceptualization: M.A.K., Å.L., F.I., A.H., M.W. and M.L.; Formal analysis: M.A.K.; Methodology: M.A.K., Å.L., F.I., A.H., M.W. and M.L.; Resources: M.A.K., Å.L., F.I., A.H., M.W. and M.L.; Validation: M.A.K.; Visualization: M.A.K.; Writing—original draft preparation: M.A.K; Writing—review and editing: Å.L., F.I., A.H., M.W. and M.L.

Funding: This research received no external funding.

Acknowledgments: The authors thank Ingrid Svedberg with colleagues at Tetra Pak Product Development Centre in Lund Sweden for assisting in the production of the UHT milk.

Conflicts of Interest: The authors declare no conflict of interest.

References

1. Ramsey, J.A.; Swartzel, K.R. Effect of ultra high temperature processing and storage conditions on rates of sedimentation and fat separation of aseptically packaged milk. *J. Food Sci.* **1984**, *49*, 257–262. [CrossRef]
2. Deeth, H.C.; Lewis, M.J. Protein stability in sterilised milk and milk products. In *Advanced Dairy Chemistry*; McSweeney, P.L.H., O'Mahony, J.A., Eds.; Springer: New York, NY, USA, 2016; pp. 247–286, ISBN 978-1-4939-2799-9.
3. Tsioulpas, A.; Lewis, M.J.; Grandison, A.S. Effect of minerals on casein micelle stability of cows' milk. *J. Dairy Res.* **2007**, *74*, 167–173. [CrossRef] [PubMed]
4. Gaucheron, F. The minerals of milk. *Reprod. Nutr. Dev.* **2005**, *45*, 473–483. [CrossRef] [PubMed]
5. Walstra, P.; Wouster, J.T.M.; Geurts, T.J. *Dairy Science and Technology*, 2nd ed.; Taylor & Francis: Boca Raton, FL, USA, 2006; ISBN 0-8247-2763-0.
6. Lewis, M.J.; Grandison, A.S.; Lin, M.-J.; Tsioulpas, A. Ion calcium and pH as predictors of stability of milk to UHT processing. *Milchwissenschaft* **2011**, *66*, 197–200.

7. Tessier, H.; Rose, D. Heat stability of casein in the presence of calcium and other salts. *J. Dairy Sci.* **1961**, *44*, 1238–1246. [CrossRef]
8. Singh, G.; Arora, S.; Sharma, G.S.; Sindhu, J.S.; Kansal, V.K.; Sangwan, R.B. Heat stability and calcium bioavailability of calcium-fortified milk. *LWT-Food Sci. Technol.* **2007**, *40*, 625–631. [CrossRef]
9. Omoarukhe, E.D.; On-Nom, N.; Grandinson, A.S.; Lewis, M.J. Effects of different calcium salts on properties of milk related to heat stability. *Int. J. Dairy Technol.* **2010**, *63*, 504–511. [CrossRef]
10. Whittier, E.O. The solubility of calcium phosphate in fresh milk. *J. Dairy Sci.* **1929**, *12*, 405–409. [CrossRef]
11. Garnsworthy, P.C.; Masson, L.L.; Lock, A.L.; Mottram, T.T. Variation of milk citrate with stage of lactation and de novo fatty acid synthesis in dairy cows. *J. Dairy Sci.* **2006**, *89*, 1604–1612. [CrossRef]
12. Faulkner, A.; Peaker, M. Reviews of the progress of dairy science-secretion of citrate into milk. *J. Dairy Res.* **1982**, *49*, 159–169. [CrossRef]
13. Carlsson, J.; Bergström, J.; Pehrson, B. Variations with breed, age, season, yield, stage of lactation and herd in the concentration of urea in bulk milk and individual cow's milk. *Acta Vet. Scand.* **1995**, *36*, 245–254. [PubMed]
14. Auldist, M.J.; Walsh, B.J.; Thomson, N.A. Seasonal and lactational influences on bovine milk composition in New Zealand. *J. Dairy Res.* **1998**, *65*, 401–411. [CrossRef] [PubMed]
15. Reid, M.; O'Donovan, M.; Elliott, C.T.; Bailey, J.S.; Watson, C.J.; Lalor, S.T.J.; Corrigan, B.; Fenelon, M.A.; Lewis, E. The effect of dietary crude protein and phosphorus on grass-fed dairy cow production, nutrient status, and milk heat stability. *J. Dairy Sci.* **2015**, *98*, 517–531. [CrossRef] [PubMed]
16. Williams, R.P.W. The relationship between the composition of milk and the properties of bulk milk products. *Aust. J. Dairy Technol.* **2002**, *57*, 30–44.
17. Crowley, S.V.; Megemont, M.; Gazi, I.; Kelly, A.L.; Huppertz, T.; O'Mahony, J.A. Heat stability of reconstituted milk protein concentrate powders. *Int. Dairy J.* **2014**, *37*, 104–110. [CrossRef]
18. Lindmark-Månsson, H. *Den Svenska Mejerimjölkens Sammansättning 2009*; The Swedish Dairy Association: Stockholm, Sweden, 2012.
19. Karlsson, M.A.; Langton, M.; Innings, F.; Malmgren, B.; Höjer, A.; Wikström, M.; Lundh, Å. Changes in stability and shelf-life of ultra-high temperature treated milk during long term storage at different temperatures. *Heliyon* **2019**, *5*, e02431. [CrossRef]
20. Hardham, J.F.; Imison, B.W.; French, H.M. Effect of homogenisation and microfluidisation on the extent of fat separation during storage of UHT milk. *Aust. J. Dairy Technol.* **2000**, *55*, 16–22.
21. Lu, C.; Wang, G.; Li, Y.; Zhang, L. Effects of homogenisation pressures on physicochemical changes in different layers of ultra-high temperature whole milk during storage. *Int. J. Dairy Technol.* **2013**, *66*, 325–332. [CrossRef]
22. Malmgren, B.; Ardö, Y.; Langton, M.; Altskär, A.; Bremer, M.G.E.G.; Dejmek, P.; Paulsson, M. Changes in proteins, physical stability and structure in directly heated UHT milk during storage at different temperatures. *Int. Dairy J.* **2017**, *71*, 60–75. [CrossRef]
23. Dalgleish, D.G. Sedimentation of casein micelles during the storage of ultra-high temperature milk products-a calculation. *J. Dairy Sci.* **1992**, *75*, 371–379. [CrossRef]
24. Van Boekel, M.a.J.S. Effect of heating on Maillard reaction in milk. *Food Chem.* **1998**, *62*, 403–414. [CrossRef]
25. Day, L.; Raynes, J.K.; Leis, A.; Liu, L.H.; Williams, R.P.W. Probing the internal and external micelle structures of differently sized casein micelles from individual cows milk by dynamic light and small-angle X-ray scattering. *Food Hydrocoll.* **2017**, *69*, 150–163. [CrossRef]
26. Malmgren, B. Long-life milk. In *Dairy Processing Handbook*; Tetra Pak Processing Systems AB: Lund, Sweden, 2018.
27. Horne, D.S. Ethanol stability and milk composition. In *Advanced Dairy Chemistry*; McSweeney, P.L.H., O'Mahony, J.A., Eds.; Springer New York: New York, NY, USA, 2016; pp. 225–246.
28. Shew, D.I. *New Monograph on UHT Milk*; International Dairy Federation: Brussels, Belgium, 1981.
29. Davies, D.T.; White, J.C.D. The stability of milk protein to heat. I. Subjective measurements of heat stability of milk. *J. Dairy Res.* **1966**, *33*, 67–81. [CrossRef]
30. Lindmark-Månsson, H.; Fondén, R.; Pettersson, H.-E. Composition of Swedish dairy milk. *Int. Dairy J.* **2003**, *13*, 409–425. [CrossRef]
31. New Zealand Dairy Industry. Fat rise-visual-for long-life milks. In *New Zealand Technical Manual. NZTM 4: Physical Methods Manual*; Fonterra Cooperative Group Ltd.: Hamilton, New Zealand, 2000; pp. 15.2.1–15.2.2.

32. New Zealand Dairy Industry. Sediment-visual "bottom cover area" for long-life milks and creams. In *New Zealand Technical Manual. NZTM 4: Physical Methods Manual*; Fonterra Cooperative Group Ltd.: Hamilton, New Zealand, 2000; pp. 15.5.1–15.5.4.
33. Karlsson, M.A.; Langton, M.; Innings, F.; Wikström, M.; Lundh, Å.S. Short communication: Variation in the composition and properties of Swedish raw milk for ultra-high-temperature processing. *J. Dairy Sci.* **2017**, *100*, 2582–2590. [CrossRef] [PubMed]
34. Zadow, J.G. The influence of pH and heat treatment on the colour and stability of ultra-high-temperature sterilized milk. *J. Dairy Res.* **1971**, *38*, 393–401. [CrossRef]
35. Zadow, J.G. The stability on milk of low pH towards UHT processing. In *Proceedings of the Proceedings of the 20th International Dairy*; International Dairy Federation: Brussels, Belgium, 1978; p. 711.
36. Gaur, V.; Schalk, J.; Anema, S.G. Sedimentation in UHT milk. *Int. Dairy J.* **2018**, *78*, 92–102. [CrossRef]
37. Wilson, H.K.; Herreid, E.O.; Whitney, R.M. Ultra centrifugation studies of milk heated to sterilization temperatures. *J. Dairy Sci.* **1960**, *43*, 165–174. [CrossRef]
38. Grewal, M.K.; Chandrapala, J.; Donkor, O.; Apostolopoulos, V.; Stojanovska, L.; Vasiljevic, T. Fourier transform infrared spectroscopy analysis of physicochemical changes in UHT milk during accelerated storage. *Int. Dairy J.* **2017**, *66*, 99–107. [CrossRef]
39. Boumpa, T.; Tsioulpas, A.; Grandison, A.S.; Lewis, M.J. Effects of phosphates and citrates on sediment formation in UHT goats' milk. *J. Dairy Res.* **2008**, *75*, 160–166. [CrossRef]
40. Jeurnink, T.J.M.; de Kruif, K.G. Calcium concentration in milk in relation to heat-stability and fouling. *Netherlands Milk Dairy J.* **1995**, *49*, 151–165.
41. Van Boekel, M.a.J.S.; Nieuwenhuijse, J.A.; Walstra, P. The heat coagulation of milk. 1. Mechanisms. *Netherlands Milk Dairy J.* **1989**, *43*, 97–127.
42. Udabage, P.; McKinnon, I.R.; Augustin, M.A. Effects of mineral salts and calcium chelating agents on the gelation of renneted skim milk. *J. Dairy Sci.* **2001**, *84*, 1569–1575. [CrossRef]
43. Miller, P.G.; Sommer, H.H. The coagulation temperature of milk as affected by pH, salts, evaporation and previous heat treatment. *J. Dairy Sci.* **1940**, *23*, 405–421. [CrossRef]
44. Muir, D.D.; Sweetsur, A.W.M. Effect of urea on the heat coagulation of the caseinate complex of skim milk. *J. Dairy Res.* **1977**, *4*, 249–257. [CrossRef]
45. Al-Saadi, M.S.J.; Deeth, C.H. Cross-linking of proteins in UHT milk during storage at different temperatures. *Aust. J. Dairy Technol.* **2008**, *63*, 93–99.
46. Gaucher, I.; Mollé, D.; Gagnaire, V.; Gaucheron, F. Effects of storage temperature on physico-chemical characteristics of semi-skimmed UHT milk. *Food Hydrocoll.* **2008**, *22*, 130–143. [CrossRef]

MDPI
St. Alban-Anlage 66
4052 Basel
Switzerland
Tel. +41 61 683 77 34
Fax +41 61 302 89 18
www.mdpi.com

Foods Editorial Office
E-mail: foods@mdpi.com
www.mdpi.com/journal/foods

www.ingramcontent.com/pod-product-compliance
Lightning Source LLC
LaVergne TN
LVHW070624170726
843515LV00004B/175

* 9 7 8 3 0 3 9 2 8 6 8 8 1 *